"We are in the midst of many and challenging environmental issues. In this outstanding and thoroughly researched work by Professor Sayan Dey, we encounter garbocracy as a form of life in contemporary India. Dey shows that garbocracy is not simply a concept but that garbo-politics is now a way through which we behave and live. This is a must-read for critical scholars of environmental justice and environmental humanities, thinking about the future of our citizenship in a more-than-human world."

– Nikoleta Zampaki, National and Kapodistrian University of Athens, Greece

"Through processes of globalization, waste has become a pervasive material presence in our lives. Sayan Dey's *Garbocracy* develops a critical ethnographic approach to understanding the cultural, social, and ecological implications of waste in India. This is an important book for those concerned with the dangers of a garbocratic future."

– John Charles Ryan, Associate Professor, Southern Cross University

"Sayan Dey takes the reader on a journey through the political, cultural, social, and intellectual politics of disposal. Surfacing the garbo-ontologies buried within the systemic injustices and violences of waste, Dey connects the social and political life of trash with the normalization of the various crises we find ourselves in today. A profoundly refreshing read in revolting times, *Garbocracy* is not to be missed!"

– Haley McEwen, University of Gothenburg

"Introducing garbageous existence as an assemblage of imperial materialities, Sayan Dey's *Garbocracy* is a timely response to the disruptive neoliberal policies that govern the knowledge-production mechanisms of human-garbage encounters."

– Basak Agin, Associate Professor, TED University, Ankara

"Trash to shit, *Garbocracy* by Sayan Dey lets us reflect on how our decomposed intellectualism may lead us to a great human collapse. The book foregrounds our dictatorially demonizing power-politics of pollution and its disposal, be it geographic, socioeconomic, optical, historical, cultural, religious, spiritual, material, sectarian, ethnic, or any other kind. It denudes our Machiavellian empires of the waste, our capitalistically gazed and cherry-picked policies, our accumulating, hoarding, and dumping mindsets."

– Waseem Anwar, Director ICPWE, Kinnaird College for Women, Lahore, Pakistan

"In response to intersecting socio-economic and socio-environmental crises, which materially impact the everyday realities of humans and non-human animals, Sayan Dey offers a way for us to re-conceptualise humans as 'garbageous entanglements.' In addressing 'the politics of trash', Dey confronts us with how we accumulate, dispose, distribute and regulate particular, neoliberal practices, emotions, and knowledges that become toxic to us and our ecosystems. He demands that we consider whether another, less wasteful world is possible, which begins from a refusal of the common sense of neoliberal norms."

– Richard Hall, Professor, De Montfort University, UK

"Sayan Dey's compelling account forces the reader to think critically about the complex historical, political and cultural dynamics of garbage, and how competing environmental concerns about waste disposal find their articulation in contemporary India."

– Avishek Ray, Cultural Historian, NIT Silchar

"Far from being a purely mundane byproduct of the global consumer culture, the accumulation, disposal, distribution and regulation of waste, trash, and filth is a multidimensional sociopolitical phenomenon characteristic of the neoliberal era, which is interwoven and implanted within us all. While democracy is being reconfigured as 'garbocracy' and human beings as 'garbo-beings', this must-read thought-provoking book written by a top-notch academic author aims at dissecting the politics and optics of garbaging in a holistic way to unravel the sickening tapestry of our 'garbocratic' future."

– Julien Paret, Director, Alliance Centre for Eurasian Studies, Alliance University

Garbocracy

Garbocracy

Towards a Great Human Collapse

Sayan Dey

PETER LANG
Chennai - Berlin - Bruxelles - Lausanne - New York - Oxford

Bibliographic information published by the Deutsche Nationalbibliothek.
The German National Library lists this publication in the German National Bibliography;
detailed bibliographic data is available on the Internet at http://dnb.d-nb.de.

A catalogue record for this book is available from the British Library.

Library of Congress Control Number: 2024051547

Cover image: 'A Plastic Burial', Sayan Dey.
Cover design by Peter Lang Group AG

ISBN 978-1-80374-729-3 (print)
ISBN 978-1-80374-727-9 (ePDF)
ISBN 978-1-80374-728-6 (ePub)
DOI 10.3726/b22308

© 2025 Peter Lang Group AG, Lausanne
Published by Peter Lang Pvt Ltd, Chennai, India
info@peterlang.com - www.peterlang.com

Sayan Dey has asserted his right under the Copyright, Designs and Patents Act, 1988, to be identified as Author of this Work.

All rights reserved.
All parts of this publication are protected by copyright.
Any utilisation outside the strict limits of the copyright law, without the permission of the publisher, is forbidden and liable to prosecution.
This applies in particular to reproductions, translations, microfilming, and storage and processing in electronic retrieval systems.

This publication has been peer reviewed.

Dedicated to…
Rotten leaves,
Dried trunks,
Dumped brains,
Infected bodies,
Silenced tongues.

Contents

(Un)Acknowledgment	ix
Foreword by Christine Daigle	xi
Why Garbocracy? Investigating Garbageous Entanglements	xv
SECTION I Accumulating	1
CHAPTER 1 Introduction: The Politics and Optics of Garbaging	3
CHAPTER 2 Garbo-Imperialism: The Empire of Wastes	19
SECTION II Disposing	29
CHAPTER 3 Garbo-knowledge: Towards Garbo-intellectualism	31
CHAPTER 4 Garbo-power and Garbo-being: The Onset of Garbo-ideologies	45

SECTION III Distributing — 57

CHAPTER 5
Garbo-pedagogies and Garbo-curricula: Schools
as Knowledge-garbage Laboratories — 59

SECTION IV Regulating — 75

CHAPTER 6
Garbo-Citizenship: The Great Human Collapse — 77

CHAPTER 7
Digital Discardscapes, Digital Infrastructures, and Garbocratic
Futures — 93

CHAPTER 8
Conclusion: Towards Counter-Garbocratic Futures — 107

Index — 125

(Un)Acknowledgment

This book has been nurtured by multiple voices, including waste collectors, rag pickers, housing residents, passersby, shopkeepers, and many more. Due to ethical and practical concerns, documenting or revealing every name may not be possible, but every voice matters. Thanks to the threatening, nauseating, fractured, and guerrilla political tactics of the currently ruling right-wing Bharatiya Janata Party (BJP) government, this project took shape. Thanks to the intergenerational caste and communal brutalities unleashed by high-caste Hindu Puritans against both humans and the natural environment, they have sharpened my observation and analytical skills and enabled me to deeply investigate their histories of garbageous egoisms. I warmly (un)acknowledge them.

Now, I acknowledge Christine Daigle, Francesca Ferrando, Marco Armiero, Nikoleta Zampaki, John Charles Ryan, Haley McEwen, Basak Agin, Waseem Anwar, Richard Hall, Avishek Ray, and Julien Paret for not only producing socially transformative and radical scholarship, but also for kindly endorsing my project. I also wish to express my warm gratitude to Madhumita Das for meticulously creating the index. The brilliant support of Indrani Dutta, Nandini Ganguli, Shruthi Maniyodath, and the proofreading and copyediting team must be acknowledged for caressing and embracing the ideas and expressions of the book with such love and care.

Foreword

In his 2012 short animation film *Man*, Steve Cutts depicts the damage that human beings have inflicted on nonhuman others and the Earth system. In the movie, a single male character wearing a "welcome" t-shirt gallivants to Edvard Grieg's *In the Hall of the Mountain King* (1875), killing nonhuman animals and destroying natural resources. He finishes his dance by climbing a trash heap of his own making, sitting triumphantly on a throne, crowning himself, and lighting a cigar. Aliens arrive on the scene, befuddled and angry at the man who has created this world of garbage; they stomp him into his piece of trash: a welcome mat they discard as they fly away from the desolate world. I have always thought this short film succinctly captures the historical and deeply problematic relationship between humans and the Earth system, especially the nonhuman others with whom they share that system. Killing and discarding for amusement's sake, overconsumption of goods and nonhuman animals, industrialization, and experimentation on nonhuman animals are on full display. The acceleration of this damaging, oppressive, and extractive activity generates more and more trash as humans inexorably move forward until the whole world is composed of garbage. The short film offers a generative platform to launch a discussion on capitalism, consumption, industrialization, the great acceleration, the Anthropocene, and the global environmental crisis; I have used it for such purposes in my courses and seminars. What the film misses, although not entirely, since the human protagonist becomes trash at the hands of the aliens, is how this "garbaging" mentality also infects how human beings relate to one another.

In our world in crisis, not only do natural resources and nonhuman animals suffer the same fate as the human protagonist of *Man*, but so do those humans toward whom slow—or sometimes even quick—violence is exerted. Such human lives are considered disposable or not considered at all. Thinking about the words we use to refer to trash is helpful here: garbage, trash, refuse, waste, and residue. There are many more, and they all have negative connotations. I like to think of "refuse" in particular, as it indicates a desire, on the part of the person who throws away, to disconnect from the junk—a refusal to engage or to be intertwined with this rejected matter. The same goes for our relations with human others. In oppressive and hierarchical societies, a garbaging attitude is often at work to a greater or lesser degree, depending on the strength of hierarchical structures and how firmly embedded the distinctions are between classes or racial groups. Even the most egalitarian societies contain distinctions, be they distinctions based on race, social or economic class, gender, sexual orientation, ability, age, or any combination thereof. When our worldview enables us to devalue and mistreat the nonhuman world, only one little step is needed to lead us to exercise the same attitude toward those humans who are deemed "below us" or simply different from us.

Thus, to think of garbage is not merely contemplating one's trash receptacle—developing strategies to fill it less with material refuse thanks to recycling, composting, or a reduction in our consumption. It is, instead, crucial to think of how our behaviors and attitudes—what and who we value in our societies and why we value them—may lead us to treat other humans as trash: humans with whom we refuse to enter into relation, humans from whom we want to keep a distance for fear of being contaminated. With what? The same dehumanization that we have imposed upon them. And yet, if we were to examine our current ways of relating with acuity, we would quickly notice that in treating other humans, nonhumans, and the Earth system in a wasteful manner, we end up dehumanizing ourselves. It is okay for the aliens to stomp on the man in the short film: he no longer deserves any respect or value because he has engaged in such relating.

Humans are deeply entangled beings. They are entangled with their fellow humans, nonhuman animals, plants, ecosystems, social and political institutions, etc. Refusing these entanglements is participating in garbocracy: the regime of wasting, garbaging, and trashing ourselves and the world. To resist garbocracy, we need to conduct a sober analysis of the current state of things, the force of existing garbocracies, and our roles in them. This attentive and reflective thinking will encourage a rediscovery of our entanglements, thus realizing that our participation in autocracies can only lead to our individual and collective demise.

Dr. Christine Daigle
Director, Posthumanism Research Institute
Brock University

Why Garbocracy? Investigating Garbageous Entanglements

"All things are meshed together ..."

(Aurelius 2006: 59)

Garbage as a physical and ideological entity has played a "valuable role" in human evolution and plays "a useful function in our current daily lives" (Nussbaum 2004: 14). The monograph *Garbocracy: Towards a Great Human Collapse* is the third part of the 'Decoloniality, Eco-sustainability, and Cultural Studies' research project that I conceived in 2018. This monograph continues *Green Academia: Towards Eco-friendly Education Systems* (published in 2022) and *Performing Memories and Weaving Archives: Creolized Cultures across the Indian Ocean* (published in 2023). The project has traversed multiple mountainways, roadways, and waterways to unfold the different patterns of garbageous existence that human civilization is entangled in. Like garbage heaps, garbageous existence is a concoction of suaveness, seduction, dictatorship, disgust, regret, arrogance, extravagance, pride, and ignorance. It is simultaneously visible and invisible, present and absent, dead and alive, and appropriative and convincing. The cohabitation of humans as garbageous entanglements is not just in terms of the physical location of specific individuals and communities within and around litter and dumping yards but also in terms of the acknowledgment and practice of garbageous ideologies, garbageous politics, and garbageous knowledge systems, which eventually have led to the evolution of garbageous beings. Garbageous beings are a community of intellectually sterilized folks who have lost their critical capabilities to think and act on their terms, either out of choice or out of compulsion. As a result, they can celebrate injustices and ethicalities in normative and ceremonious ways without self-realization and regret. This is why the raping of Dalit women is logicalized as a disciplinary initiative, mob lynching of Muslims is legalized as an act of sociocultural puritanism, murders of whistleblowers are justified as acts of protecting sociopolitical dignity, and manipulations of histories and sciences are interpreted as pious acts of reviving the Indigenous traditions of precolonial India. Garbageous existence outlines how "social incentives today gear into a paradigm of accumulation" where "going against the norms of accumulation [...] carries wider social costs" (Davis 2023). It pushes selective bodies and psyches into "inscribed surfaces" (Guru 2012: 85) that are recognized through specialized biometric identifiers like skin colours, dressing patterns, language structures, voice textures, economic conditions, and the accumulation and disposal styles of waste. The inscribed surfaces are transitional and ephemeral, where living bodies (humans and more-than-humans) are physically, psychologically, ideologically, and intellectually reduced to faecal matter—bacterial, microbial, disgusting, and nauseating.

This book was written when the constitutional rights and duties of free thinking and expression in India were compromised daily. Every paradigm of knowledge production, like curricula, pedagogies, documentation, archiving, news production, and various others, is being surveilled and regulated by right-wing neoliberal ideologies that are structurally and institutionally propagated by the Bharatiya Janata Party (BJP). The dictatorial agencies of the right-wing rule in India are maintained by a community of "infantile, stupid, fraudulent, coarse, mercenary, [and] despotic" (Clay 2006: xxi) individuals who penalize, silence, and erase any effort to generate knowledge outside the right-wing dictatorial governmentality. The arguments in this book emerge out of personal agonies, insecurities, and tensions, which I believe are being experienced by many residents in India today. Like unwanted waste materials, the toxicity of right-wing ideologies has reached such an alarming point that knowledge disciplines that do not cater to the right-wing neoliberal propaganda of the BJP are scrapped of funds and flushed out. As a result, the ritualistic practices of cancellations and erasures have reached a very normative state, where no one bothers to interrogate the illegalities and mishappenings.

This is not the first book that argues about the decolonial wrong turn that the BJP government has deliberately undertaken to satiate its communally, culturally, politically, economically, and religiously fractured neocolonial sociopolitical agendas, and, hopefully, this is not the last book. However, only a few works have dealt with the garbageous political-ecological entanglements that the governing policies of the BJP portray to justify their garbocratic designs of development and modernity. The arguments about garbage, garbaging, and garboglomerates in this book are founded on the question: Are the physical garbage that is openly, deliberately, and ignorantly disposed of in the streets and the garbageous policies and ideologies that are habitually deployed by the political organizations in India in the name of progress and smartness any different? The chapters in this book make a collaborative effort to address this question through the thematic and methodological aspects of garbo-imperialism, garbo-knowledge, garbo-power, garbo-beings, garbo-pedagogies, garbo-curricula, garbo-citizenship, garbocracy, and counter-garbocratic possibilities.

In an era when more and more we are reconfiguring our habitual existence through the phenomena of Artificial Intelligence, posthumanism, and post-anthropocentrism, it is crucial to understand our physiological ways of being and becoming not in compartmentalized terms but as cobwebbed, rhizomatic, polymerized, and entangled. It is also important to clarify here that though the arguments have vastly been drawn from diverse sociopolitical situations and intentions of the BJP rule in India, the purpose of this book is not to demonize a particular political party and its ideologies. Concerning the functional patterns of right-wing neoliberalism in India, the book invites readers to engage with more significant sociohistorical concerns, which have been so deeply embedded within the human psyche for generations that people do not feel it necessary to counter them. This book also provokes us to question urgently and fearlessly and establish human-more-than-human constellations of counter-garbocratic futures, where we can collectively voice our concerns and intimately support each other.

Works Cited

Aurelius, M. (2006). *Meditations*. Translated by M. Hammond. New York: Penguin.
Clay, D. (2006). "Introduction." In M. Aurelius (ed.), *Meditations*, pp. xi–xliv. New York: Penguin Random House UK.
Davis, B. P. (2023). "How to Embrace a Wild Fire: A Path Out of the Smoke", *Public Books*, <https://www.publicbooks.org/how-to-embrace-a-wildfire-a-path-out-of-the-smoke/?utm_content=buffer50246&utm_medium=social&utm_source=twitter.com&utm_campaign=buffer>, accessed on 16 October 2023.
Guru, G. (2012). *The Cracked Mirror: An Indian Debate on Experience and Theory*. Oxford: Oxford University Press.
Nusbaum, M. C. (2004). *Hiding from Humanity: Disgust, Shame, and the Law*. Princeton: Princeton University Press.

SECTION I

Accumulating

CHAPTER 1

Introduction: The Politics and Optics of Garbaging

"... writing about waste is a mess in itself."
(Armiero 2021: 1)

"In the era of the Anthropocene, marine life and microplastics, algae and garbage patch, zooplankton and toxic chemicals, shells and oil, among others, are merging."
(Ferrando 2023: 29)

Disgusting Encounters

What happens when we encounter piles of garbage? We feel disgusted, irritated, nauseated, and threatened. All these experiences are not just casual and momentary, but they are triggered by social, cultural, historical, political, geographical, and economic factors. These factors have been consciously or unconsciously interwoven and implanted within us and the societies around us across different moments of time and space. To elaborate further, the patterns of disposing of waste are significantly influenced by the level of socio-historical consciousness that communities have regarding hygiene. This consciousness, or lack thereof, dictates the level of accessibility to hygienic practices, the financial capability to have access to sanitary practices, the level of strategic initiatives that communities undertake to keep their localities clean, and many other aspects. When a locality is unclean, the responsibility is not just on the residents of that particular locality but also on the municipal organizations, political organizations, and the residents from the adjacent communities. For instance, dumping household wastes from neighbouring localities, nearby hospitals, and vegetable markets in my locality in Kolkata is a regular affair. The wastes are dumped by the public cleaners who work for the local municipal organization. It is important to note that the political affiliation of the current municipal body conflicts with the political affiliations of a majority of the residents in our locality. After several verbal protests, the local municipal organization took up the issue, but they did not bother to take any initiative. On the contrary, after lodging a series of complaints, the municipality chief drove us away by saying that we should first vote for their political party to expect their services.

Let us look into another example to understand further how political, cultural, and communal factors regulate garbage dumping. Since my childhood days, I have often heard people saying that the Muslim communities staying near our locality lead an unhygienic lifestyle. And, it is indeed true that the public spaces in their locality are regularly filled with piles of domestic waste. However, the reason behind such a scenario has less to do with Muslims being unhygienic and more to do with the political intentions and the strategic ignorance of the local municipality. Historically, the Muslim-occupied community that I am talking about evolved as a refugee colony in the 1970s during the Bangladesh Liberation War. Besides Muslims, there are also a few refugee Hindu residents. During the war, when the refugees escaped from Bangladesh to Kolkata, they mainly resided in the unclaimed and unoccupied lands of the city. Later, many of these lands were used to construct offices and residences or unethically used as grounds for dumping waste by the houses and the municipalities. The consistent socio-political process of invisibilization, marginalization, and stigmatization forced many refugees to regard these dumping grounds as their permanent residences.

Apart from regularizing anti-refugee socio-political attitudes, using the Muslim-dominated residential areas near my locality and other parts of Kolkata as a potential space for disposing of garbage also adds fuel to the existing narratives of anti-Muslim hatred. These instances show how a "very complex combination of cultural, social, institutional, political, organizational, ecological, historical and relational dynamics" habitually regulates the intentions and systems of disposing of wastes (Luton 1996: 4). I argue this consistent and disgusting performance of disposing of garbage across selective physical, geographical and geological spaces as the politics

Introduction: The Politics and Optics of Garbaging

and optics of garbaging. The political and optical practices of garbaging allow specific socially, culturally, politically, and economically privileged communities to gain control over others. This performance is not a momentary and ignorant act and is systematically curated by the "exploitative, garbological and fast capitalist system" (Ghosh 2021: 121). The ritual of using garbage as a tool to overpower, marginalize, stigmatize, and erase communities, institutions, and the natural environment has been historically prevalent through channelling waste from one geographical region to another. The process of marginalization and stigmatization also takes place by choking communities and the natural environment to death by strategically releasing poisonous gases from discarded chemicals, polluting the natural water bodies, artificially creating ecologically sustainable spaces, piling wastes on natural lands, and then polluting the adjacent environment by burning them; and in many other ways. This "translocationality" of the production and distribution of garbage is meticulously underlined with the factors of "race, ethnicities, colour, religious affiliation, cultural heritage, or political learnings" (Ghosh 2022: 121). These factors repeatedly despise specific communities like the refugees and Muslims through the lens of "faecal imaginary" and treat them as "unsightly" (Nixon 2013: 2) and "literally full of shit" (Dlamini 2009: 132).

Human-Garbage Cohabitation in Susuwahi, Varanasi. Photo credit: Shankhadeep Chattopadhyay.

Such an imagination also systemizes the ignorance of the municipal and political organizations in India toward specific cultural and religious communities. The aspect of ignorance is a well-established epistemological practice and is "constituted or reproduced as an aspect of power" (Feenan 2007: 514). The "epistemologies of ignorance" (Steyn 2012: 10) can be further understood through the different ways in which garbage is accumulated and disposed of by the communities in India. In the slums, townships, and other places primarily occupied by the socio-economically backward communities, the wastes are disposed of in more unorganized and unhygienic ways than in the localities inhabited by the middle-class and high-class communities.[1] While in the organized middle-class and high-class residential areas, the local municipal organizations regularly monitor the removal of domestic and public wastes; such facilities are hardly available in the socio-economically backward regions. In fact, on many occasions, the waste collected from middle-class and high-class residences is dumped in townships and slums. Dumping waste in the townships and slums is not just a casual act. It is systemically motivated by multiple forms of social, cultural, and political factors like caste biases, religious demonization, normalization of socio-economic crises, gender marginalization, etc.

These "politics of trash" (Strach & Sullivan 2023) serve as the foundation of this monograph, and the arguments in the consequent chapters unfold how the physicality and opticality of garbage invades, entangles, percolates, and mutates physically and ideologically with humans and reconfigures human beings as 'garbo-beings.' Garbological sensibilities regulate the thoughts and actions of garbo-beings. The garbological functioning of humans as garbo-beings can be located through how individuals situate their understanding of societies in the contemporary era within the materiality, spirituality, culturality, and governmentality of waste.

---

[1] Here, the terms 'middle-class' and 'high-class' have been used with respect to the socio-economic status of communities.

The Material Garbage

Often, the objects that govern the daily existence of human beings, like electronic gadgets, dresses, jewellery, showpieces, crockery, and others, function as a mere conglomeration of fancy waste materials. Without denying these objects' practical and aesthetic values, it is necessary to admit that, at times, the desire to accumulate these objects has less to do with genuine requirements and aesthetic desires and more with gaining access to elitist and economically privileged socio-cultural spaces. For instance, during personal conversations with participants, many participants revealed that they often feel an ideological compulsion to possess particular objects irrespective of their non-usage. The compulsion to possess unnecessary objects is underpinned by a "capitalist gaze" that restricts individuals to "certain time and space, to specific 'bodies,'" and "specific ways of being, seeing and saying" (McGowan 2013: 5). The lack of usage of particular objects gradually transmogrifies the objects into material garbage and the non-usable objects as material garbage eventually govern the physical, emotional and ideological existence of humans and mutate humans into garbo-beings. As garbo-beings, the identity of humans becomes no different from a heap of garbage – aimless, toxic, gross, isolative, and disgusting. Under the regulation of a capitalist gaze, the evolution of garbo-beings has led to a "mutual recognition of distinct territories and overlapping boundaries" with material garbage (Ghosh 2021: 71). However, on many occasions humans fail to realize their transition into garbo-beings because the limitless accumulation of non-usable objects is often regarded as a benchmark of spiritual progressiveness, exuberance, and novelty, where spirituality is synonymous with materialistic desirability. The phenomenon of garbo-beings will be further elaborated in the fifth chapter.

The Spiritual Garbage

The association of materialistic-spiritual value with different non-usable objects in our homes metamorphoses the material garbage into spiritual garbage through the mythically, culturally, and historically fractured attitudes of cleanliness, hygiene, possession, and dispossession. To explain further, during my stay in Varanasi between 2008 and 2018, I closely observed how the patterns of accumulating and disposing of waste in individual homes are influenced mainly by communal, caste, and religious factors. During a personal conversation, Ramesh (name changed), a 37-year-old Brahmin priest from Nagwa, Varanasi, revealed that disposing of dry wastes like flowers inside the Hindu temple premises is not unhygienic. Instead, such an act reveals the spiritual sincerity of the devotees. He said: "The dry trash disposed of inside the temple premises is hygienic and spiritually enriching. Before disposing of the trash, they are offered to the gods. So, after being offered to gods, the disposed items no longer function as trash, but attain spiritual values" (Ramesh 2022).

In India, the association of garbage with gods symbolically converts waste into a hygienic entity, and hygiene is imagined through the spiritual wisdom of cleanliness and chastity, as outlined in the various Hindu religious and mythical scriptures. For instance, in the *Atharvaveda-Veda Samhita*, it is mentioned that unclean spaces around us can be cleaned "by sweeping, *by burning*, by digging, *by the lapse of time, by the walking of cows* and by the sprinkling of water" [(italics mine) (Whitney 2018: 675)]. As a result, besides the disposal of dry waste inside the temple premises, cows are allowed to roam freely in many Hindu temples because of their holy association with Hinduism. Due to the presence of cows, the temple premises often get littered with cow dung, which is deliberately not cleaned, and, as a spiritual practice, many devotees voluntarily walk barefoot on it.

Sunita (name changed), a 25-year-old Dalit woman from Durgakund, Varanasi, complained about how high-caste households from the neighbouring areas dispose of garbage in their locality. She said: "Because we are by profession sweepers, the high-caste people believe that our only work is cleaning garbage. Therefore, people from the adjacent localities, especially the high-caste folks, openly dispose of their garbage in front of our homes and around our area without considering our hygiene" (2022). She also added that no action has been taken despite lodging several complaints with the local municipality. Instead, they are often driven away, saying that salaried public cleaners must keep the environment clean and that the municipality cannot shoulder any responsibility. This instance shows how, through the unethical disposal of garbage by the high-caste communities and the lack of any initiative by the local municipal communities in Varanasi, the existing caste-based hierarchies are reproduced, and the low-caste communities are not treated better than garbage. Like piles of waste, the low-caste communities in Varanasi and many other parts of India are systematically forced into a "repressive regime of servility in everyday life" (Shah cited in Doron and Jeffrey 2019: 97) and are "extracted, burned, pumped, emitted and flushed" (Humes 2012: 14) into the realm of invisibility.

This act of invisibilization is further systematized through people like Sarvesh (name changed), a high-caste 35-year-old man from Ramapura, Varanasi, who thinks that the low-caste community of sweepers "are spiritually, culturally and historically doomed, and so it is not wrong for them to suffer in unhygienic circumstances and clean wastes that are disposed of by others" (2022). He also added that through consistently suffering and cleaning, "they can spiritually purify themselves for a better life in the next birth" (2022). The process of using garbage as a tool to demonize and stigmatize the existence of a community consistently altogether transforms certain cultural practices into garbageous entities.

The Cultural Garbage

Cultural garbage has become "one of the accurate measures of prosperity" (Humes 2012: 11) in twenty-first-century India, especially with the onset of the right-wing BJP (Bharatiya Janata Party)-led government in 2014, which, in the name of indigeneity and decoloniality, has normalized the erasure of selective social, cultural, religious, historical, and political narratives on the one side and the distortion of selective scientific, technological, historical, and religious documents on the other. The erasures and distortions are performed by marginalizing and silencing low-caste Hindu and non-Hindu communities in the name of cultural, spiritual, and traditional revivalism; bulldozing Muslim-populated temporary settlements and slum dwellings in the name of legal rules and regulations; removing selective historical incidents from the educational curricula in the name of cultural purification; imprisoning and killing citizens for exposing the corrupt practices of the government in the name of order and discipline; and marginalizing specific critical educational disciplines as unimportant in the name of modernization, progressive and intelligent learning. Any form of resistance against such "selfish garbage" of "right-wing individualism" is systemically downplayed as collective projects of social transformation, political enlightenment, and intellectual development (Marcotte 2020).

Also, such practices may not be directly associated with garbage in the literal sense of the term. But as one transectally investigates the acts mentioned above of removals and distortions through the lens of the politics and optics of garbaging, one can understand how habitual indulgence in material and spiritual garbage leads to the growth of ideological garbage, which provokes individuals to dispose of selective ideas, paradigms, intellectual values, and knowledge disciplines as waste objects. The impact of ideological garbage can be located in the experience of Sunita: "Through cleaning filth at people's homes and public spaces, our existence, social identities, and cultural values are treated no differently from the garbage. Because of our profession, people perceive our localities and everyday lifestyles as diseased and dirty" (2022).

Based on the multitudinous rituals of accumulation and disposition, gradually, the intersectional influence of physical and ideological garbage is enabling "garbage mountains, toxic emissions, and polluted waterways" (Doron and Jeffrey 2019: 16) to dictate human beings' habitual thinking and functional patterns by regulating their physical, social, cultural, political, economic, and ideological attitudes.

Introduction: *The Politics and Optics of Garbaging*

The Governmental Garbage

Today, the dictatorial presence of garbage in our daily lives has made garbage a decision-making entity. It has generated a planetary waste empire that governs humans and other living beings. The impact of material, spiritual, and cultural garbage on human civilization makes garbage function like a governing organization, controlling our likes, dislikes, thoughts, and actions. In 2017, while doing my PhD in Varanasi, I was staying in a locality called Kabir Nagar, and one of the significant issues of that locality was the blockage of roads due to the public disposal of garbage. The municipality cleaners also did not regularly clean the garbage because they were not paid sufficient salaries. As a result, the gradual accumulation of garbage over the years impacted the existence and movement of the daily commuters, shopkeepers, and residents. A point in time came when the expansion of garbage transfigured the entire locality by forcing several makeshift shops to change their places, many permanent shops to relocate to a different place, daily commuters to take other roads to go to work, and numerous residents to leave their homes and shift somewhere else. Though the crisis was human-made, a complete topographical shift of the locality due to the expansive presence of garbage cannot be ignored. It was not only the issue of Kabir Nagar; other localities in Varanasi underwent similar experiences, where massive heaps of garbage caused road accidents, infected water pipelines, choked drainage systems, and led to the outbreak of multiple waterborne and airborne diseases.

Concerning these diverse patterns of garbaging, as discussed so far, garbage will soon become an all-encompassing and all-pervasive medium through which individuals will "make sense, stitch relationships, build lives, organize, resist, conform, and animate rage, despair and sometimes hope" in the future (Fine 2018: 13). Before discussing further, the consequences of garbaging in the following chapters, let us explore the different research methods that have been used to shape the thematic and theoretical arguments in this book.

Research Method

The arguments in this book have been extensively developed through field research and personal conversations with residents from diverse caste, class, religious, and gender backgrounds and with school students between standard eight and standard twelve in the cities of Kolkata and Varanasi. The conversations with residents have been reflected in chapters two and three, and the conversations with school students have been reflected in chapter five. The research methods that have been applied to conduct personal discussions and field research are:

a. *Walking Conversations*: The personal conversations with different research participants occurred through what Maggie O'Neill and Brian Roberts argue as "walking methods" (2020). This method, on the one hand, allowed the participants to feel free to share their experiences and, on the other hand, allowed me to relate the responses with concrete evidence. During the conversations, I walked with the participants through the roads. The narrow lanes, in between heaps of garbage, choked and overflowing drains, and stinking water bodies, not only allowed me to understand the sociopolitical patterns in which waste is disposed closely but also to investigate the relationalities of garbage disposal patterns with "everyday routines, spaces, personal and group life" of the participants (O'Neill & Roberts 2020: 2).

b. *Critical Bifocality*: The information generated by walking conversations has been moulded and presented through "critical bifocality" (Weis & Fine 2012: 173) in this book. Critical bifocality can be understood "as a way to think about epistemology, design and the politics of … research, as a theory of method in which researchers try to make visible the sinewy linkages or circuits through which structural conditions are enacted in policy and institutions, as well as how such conditions come to be woven into community relationships and metabolized by individuals" (Weis & Fine 173). Besides listening to the conversations and simultaneously experiencing the stink and the disgust of garbage around me, the arguments around intersubjective political and optical practices of garbaging in India, about government policies, municipality initiatives, and community consciousness have gained further theoretical and methodological momentum through critical bifocality. This method also unpacks how, through garbaging, "difference is constructed and used as a political tool" (Kobayashi 1994: 78).

Based on these research methods, the book has been thematically divided into four sections and eight chapters.

Chapter Summaries

The first section, titled "Accumulating," consists of two chapters explaining the theories, methodologies, contexts, and sociopolitical positionalities through which the arguments in this book have been shaped. The first chapter, "Introduction: The Politics and Optics of Garbaging," sets the book's pace by introducing this work's theoretical and methodological intentions. The chapter also engages with my positionalities and the contexts from where the arguments in this book emerge. To flesh out the overarching thematic aspects of this book, the chapter talks about material, spiritual, political, and governmental garbage and how it evolves, accumulates, operates, and suffocates humans and more-than-humans.

The theories and methodologies around the patterns of accumulating garbage have been further problematized in the second chapter titled "Garbo-Imperialism: The Empire of Wastes." This chapter discusses how garbage in India functions as an imperial entity by dictating and regulating our habitual existential patterns physically and cognitively. While unfolding the different imperialistic dimensions, the chapter engages with the ethics and politics in which garbage is disposed of in Kolkata and Varanasi. I have mainly selected these geo-cultural-political locations because Kolkata is my place of birth. Varanasi is where I have resided, studied, researched, and undergone many socio-emotional experiences for over a decade. The chapter also discusses how garbo-imperialism impacts the economic status of the communities in India and how the accumulation and disposal patterns.

The second section, titled "Disposing," consists of two chapters that analyze how garbo-imperialism operates through the paradigms of knowledge, power, and being. The third chapter, "Garbo-knowledge: Towards Garbo-intellectualism," engages with how the visible and invisible presence of garbage around us impacts our habitual ways of knowledge production and our relationships with diverse caste, class, religious, racial, and gendered communities in India. Apart from influencing relationships between individuals and communities, the patterns of accumulating and disposing of garbage in India also shape different sociopolitical ideologies, power structures, vote banks, and the functional procedures of governing organizations. This chapter also talks about how the parameters of garbo-imperialism and garbo-intellectualism are entangled and interwoven. To elaborate a bit further, when the ideologies of garbo-imperialism invade the psyches, emotions, and habitual beliefs of the people, the evolution of garbage intellects and garbo-intellectualism takes place.

These aspects have been further discussed in the fourth chapter, "Garbo-power and Garbo-being: The Onset of Garbo-ideologies." The phenomena of garbo-knowledge and garbo-power, in turn, lead to the formation of 'garbo-beings.' As garbo-beings, the human entities ideologically transmute into posthuman entities. The behavioural, physical, intellectual, and cognitive functions of the posthuman self are not exclusively controlled by the whims and fancies of the human psyche but are widely influenced by the enormous physical and ideological garbage that is generated daily by digital objects, social media, artificial intelligence, socioeconomic conflicts, and political propaganda. The phenomenon of garbo-ideologies (both material and immaterial) has been discussed mainly in the sociopolitical contexts of the right-wing BJP government and how, under their governing regime, garbage-like mindsets, knowledge systems, body languages, and cultural attitudes are problematically celebrated as traditional and authentic to India.

The third section, titled "Distributing," consists of one chapter. The chapter outlines how the phenomena of garbo-knowledge, garbo-power, and garbo-being are systemically, epistemically, institutionally, and ontologically produced in India. This argument has been justified in the context of school curricula and pedagogies. The chapter "Garbo-pedagogies and Garbo-curricula: Schools as Knowledge-garbage laboratories" talks about how through endless degrees, certificates, awards, and medals, the learners and the teachers in the schools are consistently encouraged to embrace impractical, uncaring, unsustainable, and garbage modes of teaching–learning procedures that mostly lead to the production of physical and emotional wastes like uncontrolled materialistic desires, arrogance, cultural hatred, ecological imbalance, and others. This chapter investigates the various rooted patterns of producing garbo-ideologies through curricular and pedagogical transformations. The chapter also analyses how schools and higher educational institutions systemically facilitate such processes.

The fourth section is titled "Regulating" and consists of three chapters. The chapters "Garbo-citizens: The Great Human Collapse," "Digital Discardscapes and Digital Infrastructures in Urban India," and "Conclusion: Towards Counter-Garbocratic Futures" engage with how the different political and optical aspects of accumulating, disposing of, and distributing garbage in the physical forms as waste mountains and in the ideological forms as knowledge, power, being, curricula, pedagogies, and reckless digitization have led to the evolution of a class of garbo-citizens in India – a disturbing, distorted, self-centred, ignorant, capitalistic and pseudo-intellectual community of culturally and sociologically brainwashed characters. Being a garbo-citizen has become a norm, a status quo, and an underlying identity of success, smartness, growth, and development in contemporary India. While discussing garbo-citizenship, these chapters also explain how this project generally relates to my previous monographs of *Green Academia* (2022) and *Performing Memories and Weaving Archives* (2023). The garbo-citizens are gradually steering the country towards a 'garbocratic' future. Despite these challenges, the final chapter has explored the possibilities of building counter-garbocratic futures through multirooted symbiocracies and archipelagic solidarities. Some of the symbiocratic practices discussed in these chapters are Give Me Trees (a pan-Indian environmental conservation project), Garbage Café of India, and plasticulture.

Altogether, the arguments and discussions in this book call for "a rethink of the body – human, animal, Humanimal, plant – and thereby the ontological status of species too" (Nayar 2023: 8). The different paradigms of garbaging have been elaborately reflected in the next chapter through various contextual instances and personal conversations. The chapter also discusses how the phenomenon of garbaging is maintained through invisible garbage contracts.

Works Cited

Armiero, M. (2021). *Wasteocene*. Cambridge: Cambridge University Press.
Dlamini, J. (2009). *Native Nostalgia*. Auckland Park: Jacana Media.
Doron, A. & Jeffrey, R. (2019). *Waste of a Nation: Garbage and Growth in India*. Cambridge, Massachusetts and London: Harvard University Press.
Feenan, D. (2007). "Understanding Disadvantage Partly through an Epistemology of Ignorance", *Social & Legal Studies*, 16 (5), 509–531.
Ferrando, F. (2023). *The Art of Being Posthuman in the 21st Century: Who Are We in the 21st Century?* Cambridge: Polity Press.
Fine, M. (2018). *Just Research in Contentious Times: Widening the Methodological Imagination*. New York: Teachers College Press.
Ghosh, R. (2021). "The Plastic Turn", *Diacritics*, 49 (1), 64–85.
Humes, E. (2012). *Garbology: Our Dirty Love Affair with Trash*. New York: Avery.
Kobayashi, A. (1994). "Coloring the Field: Gender, 'Race,' and the Politics of Fieldwork", *Professional Geographer*, 46 (1), 73–80.
Luton, L. S. (1996). *The Politics of Garbage: A Community Perspective on Solid Waste Policy*. Pittsburg: University of Pittsburg Press.
Marcotte, A. (2020). "The Pandemic Exposes the Truth: Right-wing 'Individualism'", *Salon*, <https://www.salon.com/2020/05/14/the-pandemic-exposes-the-truth-right-wing-individualism-is-just-selfish-garbage/>, accessed on 11 January 2022.
McGowan, T. (2013). "The Capitalist Gaze", *Discourse*, 35 (1), 3–23.
Nayar, P. K. (2023). "Looking through the Symbiotic Lens". In P. Karpouzou and N. Zampaki (eds.), *Symbiotic Posthumanist Ecologies in Western Literature, Philosophy and Art: Towards Theory and Practice*, pp. 7–8. Oxford: Peter Lang.
Nixon, R. (2013). *Slow Violence and the Environmentalism of the Poor*. Massachusetts: Harvard University Press.

O'Neill, M. & Roberts, B. (2020). *Walking Methods: Research on the Move*. London & New York: Routledge.
Ramesh. (2022). "About Disposing of Garbage inside the Hindu Temples", Personal Conversation, Varanasi.
Sarvesh. (2022). "About the Relation Between Religion and Garbage", Personal Conversation, Varanasi.
Steyn, M. (2012). "The Ignorance Contract: Recollections of Apartheid Childhoods and the Construction of Epistemologies of Ignorance", *Identities: Global Studies in Culture and Power*, 19 (1), 8–25.
Strach, P & Sullivan, K. S. (2023). *The Politics of Trash: How Governments Used Corruption to Clean Cities, 1890–1929*. New York: Cornell University Press.
Sunita. (2022). "About Disposing Garbage in the Dalit Localities", Personal Conversation, Varanasi.
Weis, L. & Fine, M. (2012). "Critical Bifocality and Circuits of Privilege: Expanding Critical Ethnographic Theory and Design", *Harvard Educational Review*, 82 (2), 173–201.
Whitney, W. D. (2018). *Atharva-Veda Samhita, Vol. 2 of 2: Translated with a Critical and Exegetical Commentary; Books VIII and XIX, Pages 471–1046*. London: Forgotten Books.

CHAPTER 2

Garbo-Imperialism: The Empire of Wastes

> "Garbage Mountain was born."
> (Humes 2012: 77)

> "Right now, the earth is full of refugees, human and not, without refuge."
> (Haraway 2015: 160)

Kingdoms of Trash

The kingdoms of trash are mobile, immobile, seductive, manipulative, cognitive, visible, invisible, authoritative, and pervasive. The kingdoms evolved from tiny plastic bags or trash cans, which gradually accumulated into multiple heaps of garbage. While travelling between Varanasi and Kolkata, the train often stopped at the Kashi station due to traffic congestion. Beside the station was an open field that was unethically converted into a dump yard by the residents, and close to the yard, there was a temporary settlement of public cleaners. Every time the train stopped, I looked out the window and noticed that the mountains of garbage had multiplied from the previous time. Besides the garbage mountains, I also noticed how children, men, and women climbed and walked across the mountains as if they were out for a treasure hunt – laughing, giggling, fighting, competing, and celebrating. During my ten years in Varanasi, I gradually noted how the open field was eventually covered with garbage mountains, some even taller than the nearby double-storey houses.

In 2018, the last time I passed by the mountains of trash, I noticed that the mountains had reached enormous heights, and the houses beside the field were no longer visible from the train. The trash mountains did not appear to me that day; they were just deadly heaps of trash. Still, a living authoritarian entity is gradually crawling, invading, and occupying the physical, ideological, and cognitive existential spaces of the humans around. I felt disgusted by the sight that day and quite intimidated. My mind was flooded with several questions: What would happen if these garbage mountains kept increasing? What happens if the mountains collapse on the cleaners and the ragpickers who walk through the mountains daily to collect tins, plastics, and other dry wastes? What happens if poisonous gases are suddenly leaked from the trash? Throughout my journey, I was troubled by these questions. Physically, I was far away from those trash mountains, but the tormenting images of the trash and the people walking on it kept haunting me. Even after reaching Kolkata, it felt as if parts of the trash mountain had been embedded within me, and I was physically and psychologically girdled by waste.

Such an experience outlines that the act of perceiving garbage being thrown away and put to a "final resting place" (Nagle 2013: 26) is "emphatic and vague" (Nagle 32). Garbage is not thrown away, but in the act of throwing away, wastes are consciously and unconsciously distributed and spatialized across residential areas and natural environments. Though we imagine that we are disposing of garbage far from us, somatically and sensually, we get trapped in it. This ontological impact of garbage has reconfigured the non-functional assemblage of discarded objects into a hyper-functional imperialistic entity and made us realize that "garbage Is, always. We will die, civilization will crumble, and life as we know it will cease to exist, but trash will endure. There it was on the street, our ceaselessly erected, ceaselessly broken cenotaphs to ephemera and disconnection and unquenchable want" (Nagle 50). As a kingdom, trash provokes us to think: Who is trying to chase whom? Are the humans chasing away trash, or are trashes chasing away humans? The relevance of this question is further captured in the lamentation of Gopal (name changed), a 24-year-old male ragpicker from Belgachia,[1] Kolkata: "No matter how much we burn away the trash, it keeps on increasing. One evening, we burn the trash dumped

---

1   A locality in North Kolkata.

by the passersby around our houses, and by the time we wake up the next morning, a new heap of trash lies around us" (2022). He also adds: "At times, I feel like running away to a far-off place, far away from the stinks and the diseases. But, the next moment, I am reminded that I need to feed my family, and who else can feed me except these *heaps of garbage*?" [(2022), italics added]. Rajib's job is to separate the wet and dry garbage before the sweeper collects it, and in return, he gets paid daily.

The imperialistic and regulatory impacts of garbage on the habitual socio-economic existence of the people in India are reflected in the experiences of Rajib (name changed) as well. Rajib, a 32-year-old male from the Shyambazar region of Kolkata, who works as a clerk in a government office, observes: "Earlier, I could easily walk to my office, but now with the trash and overflowing drains in my locality, I can barely walk. As a result, I have to take an autorickshaw every day to cover a very short distance, and my monthly expenditures have drastically increased" (2023). Doubtlessly, the open disposal of wastes and overflowing drains are caused by humans. Still, a point of time comes when, due to consistent ignorance, the whole situation goes out of control. Despite initiatives like daily collection and processing of waste, garbage volumes keep increasing.

Besides ignorance, the imperialistic character of garbage has also been shaped by spatializing the disposal process in definite socio-political ways. The following section elaborates on spatialization through the algorithms in which waste is disposed of and garbagescapes generated across residential and commuting areas.

The Spatiality of Garbage Kingdoms: The Algorithms of Garbagescapes

The politics and optics of garbaging discussed in the previous chapter invite us to think further about the social, cultural, communal, and religious algorithms that keep certain localities clean at the cost of littering other localities. To elaborate further, the volume of garbage varies from one locality to another. While I was staying in Nagwa, every day, I observed that an aged man who remained in the house opposite mine would stealthily throw his plastic garbage bag at the door of my home. When I informed my house owner, they asked me to ignore him. However, after a certain period, this daily experience became intolerable, and I verbally confronted the man one day. To my utter surprise, he was not regretful of his act. Instead, he justified it by saying it was his way of "purifying himself and the locality" (2022). The man, known by the name Kaushal (name changed), said: "The house where you are staying belongs to a dirty low-caste family, and they need to be daily reminded of the place they belong to. I throw garbage at them because that is what they are" (2022). During my stay in that locality, I was surprised to see that, except for me, no one was bothered by this act.

A point in time came when the entrance to my house appeared like a dump yard, and the disgusting stink and the visuals of the accumulated waste started bothering everybody in the locality, including Kaushal. But did that stop Kaushal from disposing of garbage in front of my house? No, not at all. On the one hand, he would shout at the public cleaners and ask them to do their cleaning job sincerely; on the other hand, he would continue throwing his household waste into the road. The garbage accumulation in that locality became so severe that many people, especially boarders like me, started shifting to cleaner localities. Besides houses, there were also many student hostels around, and due to the increasingly unhygienic situation of the locality, they also began losing boarders. Treating low-caste and outcast communities as discarded objects is not new in India (Sharma 2014; Mohanty & Dwivedi 2018; Harriss-White 2020; Sur 2020). But does that make the upper castes and the privileged classes less discarded? When garbage accumulates in a locality, does it affect the privileged communities less? At least my experiences in Varanasi and Kolkata don't suggest so. When the kingdom of waste evolves, it invades, dominates, and dictates the existential patterns of humans in unbiased ways.

When I shifted to Durgakund from Nagwa, the garbagescape algorithm was less caste-based and more religion-based. Before discussing the "circuits of waste production" (Harriss-White 2020: 241), it is important to sketch the residential map of Durgakund and how different class and religious communities are located.

Garbo-Imperialism: The Empire of Wastes

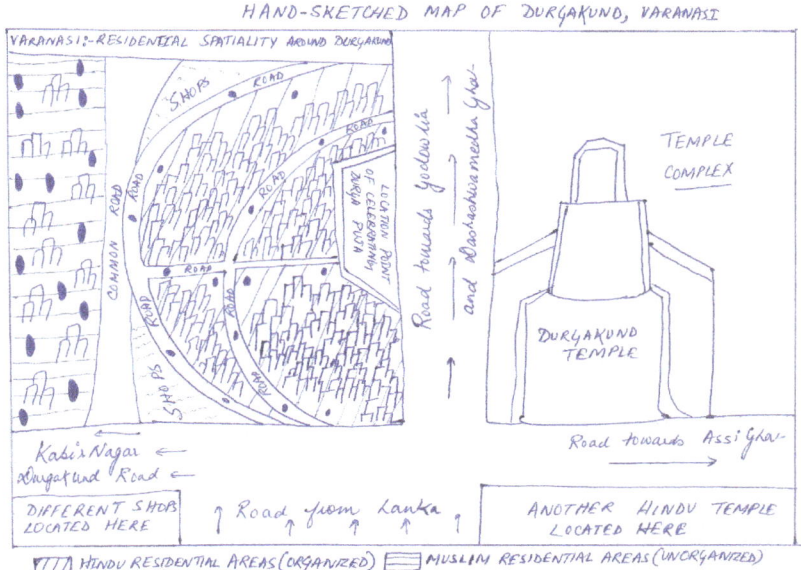

Hand-sketched map of Durgakund, Varanasi, as perceived in September 2022.[2]

In the map, it can be observed that the Hindu communities are mainly located near the Hindu temples in Durgakund. In contrast, the Muslim communities are physically located far from the Hindu residencies and their temple complexes. It can also be seen in the map that the different physical locations of the Hindu and Muslim communities impact the public places where garbage is systemically disposed of. As highlighted in the map, garbage disposal in and around the Hindu residential areas is much lower than in the Muslim residential areas in Durgakund.

An identical situation can also be seen in Belgachia, where the volume of garbage is higher in the Muslim-dominated residential areas compared to the Hindu-dominated regions. As mentioned in the previous chapter, a

[2] In September 2022, as part of my field research for this project, I visited Varanasi for seven days, and the map has been structured based on walking, observing, and communicating with people in Durgakund and Kabir Nagar.

lot of the trash dumped publicly in and around the Muslim residential areas is collected from adjacent localities. Due to different patterns of disposing and accumulating garbage in other regions, the kingdoms of waste also socially, culturally, and emotionally operate at various intensities. Najeeb (name changed) and Ravi (name changed) are colleagues at a government healthcare centre in Varanasi, and both stay in Kabir Nagar. Najeeb stays in a Muslim residential area, and Ravi in a Hindu residential area. Usually, they travel together to the office on their respective bikes and through the Muslim residential area. During a conversation, Ravi and Najeeb shared how disgusted they feel about passing through the area daily. As Ravi said, "When I pass through the area, I hold my breath. The horrible sight and the stinking smell make me feel like the garbage heaps are trying to captivate me and suffocate me to death" (2023). Najeeb adds further: "My condition is worse than Ravi's. At least he does not have to live in the garbage. But, day and night, my family and I must live with this ridiculous stink and sight. We not only feel physically unhealthy, but psychologically suffocated and traumatized as well" (2023).

Such socially, culturally, and religiously motivated unequal disposal of wastes has given rise to unequally spatialized garbage kingdoms, which are a "kind of archipelago – patchy, uneven, and not necessarily coherent" (Lepawsky 2018: 15). The narratives of the participants also show that the kingdoms are physical as well as cognitive in nature where specific communities are forced to "acquire a porous interchangeability with pollution" (Patranobish 2023: 42). As, on one side, the "unmarked" (Brekhus 1998: 35) heaps of garbage in India generate multiple physical health concerns, the psychological and neurological effects of waste on the other side cannot be ignored. The spatial and ontological articulations of waste kingdoms conceptualize "geographies of waste as enabling forms of situated knowledge" and sustain "provisional place-based idioms of subjectivity, community, and coexistence" (Patranobish 39). As Romila (name changed), a 50-year-old municipality cleaner from the Shobhabazar region of Kolkata observes: "Living in and around wastes throughout my life makes me feel depressed. It always feels like there is no life, no world for me beyond the heaps of garbage. My destiny is stuck within cleaning, burning, and processing garbage. I can barely breathe. At times, I feel like dying" (2022). She also adds: "My father was also a municipality cleaner, and he is suffering from several physical and psychological issues. Is our condition anything different from the heaps of the garbage around us?" (2022). Shyamal's experience is no different. Shyamal (name changed) is a 42-year-old resident in the Paikpara region of Kolkata who lives in a housing complex. But that does not allow him to escape the menaces of garbage kingdoms. His apartment is near the complex's main gate, and he observes: "Every day, I walk out of the complex gate to find that the road is littered with solid and liquid wastes disposed of by the houses around. The terrible sight and smell not only bother me when I am out on the road, but also when I am inside the house, and it makes me feel sick all the time" (2023). He also says: "My movements inside the house are regulated by the wastes outside. I cannot stand on the balcony and look outside because the sight of the road is so disgusting. I have to keep some of the windows permanently closed in the house because of the toxic smell" (2023).

The sociopolitical evolution and the impact of the kingdoms of waste outline how a strategically generated crisis by the ignorant public and corrupted municipalities for "designating and disenfranchising certain populations" (Z. Bauman 2004: 86) pushes every human community into a state of "existential immobility or stuckedness" (Hage 2009: 464) with garbage. The experiences of stuckedness are further entrenched within the human body and the psyche through an invisible garbage contract.

The Garbage Contract

The phenomenon of the garbage contract can be understood as a complex, invisible, and abusive sociohistorical agreement between commercial organizations, political institutions, caste, class, race, religion, mindsets, attitudes, lifestyles, and wastes that supports the simultaneous presence and absence of garbage in selective public places and gives birth to economically transactional relationships between humans and garbage. The relationships are systemically, epistemically, and ontologically maintained. Furthermore, on the one hand, the contract allows specific privileged communities to continue with their exploitative practices in society and socioeconomically underprivileged communities to become financially self-dependent at the cost of submitting, suffering, suffocating, and dying; on the other hand, it gradually allows wastes to regulate the functions and the transactions. The garbage contract is generated through "assemblages of organic species and abiotic actors" and is underlined with "extraordinary burdens of toxic chemistry, mining, depletion of lakes and rivers under and above ground, ecosystem simplification, vast genocides of people and other critters, etc., etc., in systemically linked patterns that threaten major system collapse after major system collapse after major system collapse" (Haraway 2015: 159). The contract also creates a "state of indecisive action" (Robinson 2012: 32) which is underlined with mutual performances of casualness, insincerity, ignorance, derision and hierarchies.

In my residential area in Kolkata, large machines are often deployed to clean the massive and stinking heaps of waste from the road. However, the cleaning process is primarily unenthusiastic and hasty, involving picking up some of the waste and leaving behind some. At times, the waste is picked up in such a manner by the machines and put into the garbage trucks that the road appears to be more littered than cleaned. Once, I asked a truck driver named Rajesh (name changed) what the value of such a cleaning process is when so much waste is left behind. Rajesh, who identifies himself as a high-caste Hindu, responded that I should not be bothered because the municipality cleaners and ragpickers "will take care of the rest as they must do so" (2023). Rajesh said this to me in front of a few cleaners, who enthusiastically agreed with Rajesh and nodded in consent. They said that they were standing there waiting for the trucks and machines to leave so that they could clean the remaining garbage. They also shared that these extra cleaning jobs fetch them extra payments. Though, in the name of histories, cultures, traditions, and rituals, specific communities constantly dictate, demonize, dehumanize, and discard the existential patterns of other communities as garbage, isn't garbage, as a dictatorial entity, bulldozing the "god-like omniscience and rationalist mastery" of human beings towards "low, abject, minor, diminutive registers of being and action"? (Patranobish 2023: 43). Aren't the diverse patterns of disposing and cleaning garbage in

India provoking us to think that the formation of physical garbage around us, besides caste, class, and religious prejudices, is widely a reflection of garbageous intellectual growth triggered by fictionalized scholarships on histories, philosophies, and sciences?

Works Cited

Bauman, Z. (2004). *Wasted Lives: Modernity and Its Outcasts.* Oxford, New York & Boston: Polity Press.
Brekhus, W. (1998). "A Sociology of the Unmarked: Redirecting Our Focus", *Sociological Theory*, 16 (1), 34–51.
Gopal. (2023). "About the Controlling Impact Wastes on Humans", Personal Conversation, Belgachia.
Hage, G. (2009). "Waiting Out of the Crisis: On Stuckedness and Governmentality", *Anthropological Theory*, 5 (1), 463–475.
Haraway, D. (2015). "Anthropocene, Capitalocene, Plantationocene, Chthulucene: Making Kin", *Environmental Humanities*, 6, 159–165.
Harriss-White, B. (2020). "Waste, Social Order, and Physical Order in Small-Town India", *Development Studies*, 56 (2), 239–258.
Humes, E. (2012). *Garbology: Our Dirty Love Affair with Trash.* New York: Avery.
Kaushal. (2016). "About Disposal of Wastes on the Roads", Personal Conversation, Nagwa.
Lepawsky, J. (2018). *Reassembling Rubbish: Worlding Electronic Waste.* Massachusetts: MIT Press.
Mohanty, R. & Dwivedi, A. (2018). "What Would Urban Sanitation Look Like Without Caste?", *The Wire*, <https://thewire.in/caste/what-would-urban-sanitation-look-like-without-caste>, accessed on 24 June 2023.
Nagle, R. (2013). *Picking Up: On the Streets and Behind the Trucks with the Sanitation Workers of New York City.* New York: Farrar, Strauss and Giroux.
Najeeb. (2022). "About Living Amidst Garbage", Personal Conversation, Kabir Nagar.
Patranobish, P. (2023) "Discard as Extractive Zone in Chen Quifan's *Waste Tide*", *SFRA Review*, 53 (2), 38–46.
Rajesh. (2023). "About Cleaning Garbage from Public Roads", Personal Conversation, Paikpara.

Rajib. (2023). "About the Controlling Impact of Wastes on Humans", Personal Conversation, Shyambazar.
Ravi. (2022). "About Living Amidst Garbage", Personal Conversation, Kabir Nagar.
Robinson, K. S. (2012). *2312*. London: Orbit.
Romila. (2022). "About Physical and Psychological Impact of Garbage", Personal Conversation, Shobhabazar.
Sharma, B. (2014). "India Lower Caste Still Removing Human Waste", *Al Jazeera*, <https://www.aljazeera.com/features/2014/8/25/india-lower-caste-still-removing-human-waste>, accessed on 18 July 2023.
Shyamal. (2023). "About Physical and Psychological Impact of Garbage", Personal Conversation, Paikpara.
Sur, P. (2020). "Under India's Caste System, Dalits Are Considered Untouchable. The Coronavirus Is Intensifying That Slur", *CNN*, <https://edition.cnn.com/2020/04/15/asia/india-coronavirus-lower-castes-hnk-intl/index.html>, accessed on 31 August 2023.

SECTION II
Disposing

CHAPTER 3

Garbo-knowledge: Towards Garbo-intellectualism

> "The production of 'human waste', or more correctly wasted humans (the 'excessive' and 'redundant', that is, the population of those who either could not or were not wished to be recognized or allowed to stay), is an inevitable outcome of modernization, and an inseparable accompaniment of modernity."
>
> (Z. Bauman 2004: 5)

Decomposed Intellectuality

Who is a decomposed intellectual? What relationalities do decomposed intellectuality, discarded objects, and a decomposed intellectual share? These questions are motivated by Zygmunt Bauman's concept of wasted humans and serve as entry points to this chapter. Just like garbage, decomposed intellectuality is a result of socio-politically rotten perspectives and ideologies that have steadily accumulated across generations. The causal factors of waste production discussed in previous chapters require deeper deliberation through the lens of redundant knowledge production. This chapter explores how the processes of waste production and redundant knowledge production in India are correlated, i.e., waste produces non-essential knowledge, and non-essential knowledge produces waste. But how? The material objects we use daily have "histories arising from everyday personal rituals" (Doron and Jeffrey 2019: 150). These histories and rituals not only relate to the use of different objects but also extend to the structures of waste disposal. As shared in the previous chapter, Kaushal's daily practice of throwing garbage in Nagwa or Najeeb's experiences of living amidst garbage in Kabir Nagar are outcomes of decomposed intellectualities produced by caste-based and religion-based divisions. These decomposed intellectualities ultimately result in "mutually assured vulnerability" (Z. Bauman 7). This mutual vulnerability is starkly reflected in the experiences of Naveen, an 18-year-old rag picker in the Belgachia region. He observes:

> Every day, I see people throw plastic bags and garbage around our homes and in our localities while they pass. They will pass through our locality with masks and handkerchiefs across their nose to fight the disgusting smell. But, still, they continue to throw garbage at us because we are destined to be treated like garbage. (2023)

Naveen also raises a crucial question: "Isn't this problematic attitude affecting all of us?" (2023). When he tries to stop people from dumping waste in his locality and asks this question, it is met with blank smiles, confused expressions, abuse, and more waste the next day. Despite the awareness of catastrophic consequences, the daily engagement with such revulsive acts becomes systemically possible through the symbolic replication of disgusting discards and detestable intellectualities, which Josh Lepawsky identifies as "worlding" (2018: 5). According to Lepawsky, worlding is a "jumble of not necessary like things – people as citizens, consumers, and corporations; materials such as plastics, metals, and glass; energy and information; sites and situations" that become connected through waste, cohering "to form a common world" (2018: 6).

The everyday entanglement of waste and wasted humans shares a "semantic space with 'rejects', 'wastrels', 'garbage', [and] 'refuse'" and they are disposed of "like the empty, non-refundable plastic bottle or once-used syringe – an unattractive commodity with no buyers, or a substandard or stained product" (Z. Bauman 12). Like waste, many human bodies, psyches, and intellectualities in India are rejected, erased, and disposed of, making them "invisible by not looking and unthinkable by not thinking" (Z. Bauman 27). In general, the modern capitalist society in India is "synthetic [and] unpredictable" (MacBride 2012: 174), which, in the name of heterogeneity, fosters diverse forms of sociopolitical and intellectual toxicity (Gray-Cosgrove, Liboiron & Lepawsky 2015). The phenomenon of human-waste entanglement is further discussed in the following sections, examining the peculiar pseudo-decolonial projects of re-indigenizing India by the BJP government, which recycle stories of "cosmic fear ... into the 'official' kind" (Z. Bauman 48, also see Bakhtin 1968). To contextualize the communions between wasted humans and garbage, I specifically refer to decolonization (or pseudo-decolonization!) because the BJP government treats the Muslims, Dalits and refugees in India as if they were waste, by unleashing ethnic violence, distorting scientific and cultural facts, and erasing intellectual diversity in their decolonial acts of restoring the 'golden' period of ancient India (Special Correspondent 2022; ANI 2022; PTI 2023).

Garbo-science

The human-waste entanglement, besides physical wastes, has given birth to "'affective' (feelings/emotions), 'behavioural' (intentions/actions), 'cognitive' (knowledge/beliefs) and 'evaluative' (values/likes or dislikes)" (Teo and Loosemore 2010: 742) garbage. These forms of waste are evident through the immense production and accumulation of garbage science in India, derived from Hindu mythologies and embellished with ceaseless imaginations. Let us examine some of the garbo-scientific narratives that the fake laboratories of the Bharatiya Janata Party (BJP), the current right-wing ruling party of India, have been propagating so far. In 2014, Narendra Modi, the Prime Minister of India, at an Indian Science Congress meeting, citing examples of Hindu mythological figures Ganesha and Karna, claimed that "reproductive genetics and cosmetic surgery" existed "thousands of years ago" in India (Modi quoted in Rahman 2014). In the same year, while addressing a group of school students, Modi declared that there is nothing called climate change. People "lose their ability to tolerate the cold as they grow older" (Modi quoted in Narayanan 2014). In 2015, without any scientific evidence, Maneka Gandhi, an animal rights activist and a member of the BJP, proposed the use of cow urine as a floor disinfectant instead of chemical phenyls (Scroll Staff 2015). In 2017, the education minister of Rajasthan (a state in western India), Vasudev Devnani, argued that cows inhale and exhale oxygen, so diseases like cough, flu, and cold can be cured by "just going near it" (Devnani quoted in Vyas 2017). In 2018, the Union Minister of India Satyapal Singh philosophized that Charles Darwin's theory of man's evolution is "scientifically wrong" because "nobody, including our ancestors, in written or oral, has said they saw an ape turning into a man." (Singh quoted in Scroll Staff 2018).

These pseudo-scientific dimensions have been verbally increased and deployed systemically and institutionally through curricula and pedagogies in schools and higher educational institutions in India. What is more alarming is that such problematic arguments have been strategically appreciated in the Euro-North American-centric West as a decolonial move, which is not the reality. The physical, ideological, and intellectual crises that these garbo-scientific knowledge paradigms have generated, on one side, demean, distort, and dismiss the genuine sciences and technologies of precolonial India, and on the other side, ensure the preservation and domination of western sciences. Therefore, the West's appreciation of these garbage scientific narratives should make us sceptical and provoke us to ask – is this appreciation a neocolonial mockery in disguise?

It is important to clarify here that the purpose of critiquing the attitude of deriving scientific arguments from Hindu mythologies is not to demean the social, cultural, and spiritual values of Hindu mythologies. There is concrete and relevant evidence that India had its infrastructure for science, engineering, and technology studies before the European invasion. For example, the three-dimensional architectures, water ducts, astronomical science, accurate geometrical shapes, and the stone carvings of drinking straws, rocket artilleries, and telescopes in the Chennakeshava Temple and the Hoysaleshwara Temple reveal the existence of local Indigenous sciences, technologies, and mathematics in India during the 10th century. However, while engaging with the aspects of Indigenous sciences, this evidence is mainly ignored, and the discussions, as mentioned above, are mostly centred on non-evidential or anti-evidential aspects.

It is the pseudo-scientific practices that function as the "normative science of the conduct of human beings living in societies" (Lillie 1957: 2) and have led to the development of "deleterious social consequences" (Wacquant 2001: 40), like lack of logical and critical interpretations, incessant breeding of cancel cultures, rise of health and medical crises, distortions of curricular

Garbo-knowledge: Towards Garbo-intellectualism 35

and pedagogical practices in educational institutions, and many others. The sociopolitical rituals of "disengagement, discontinuity and forgetting" (Z. Bauman 117) are also performed through the imagination and production of garbage philosophies.

Hoysaleshwara Temple – Rocket Artillery. Photo credit: Sayan Dey.

Hoysaleshwara Temple – Straw and Telescope. Photo credit: Sayan Dey.

Garbo-philosophy

Garbo-philosophy "entails an unavoidable – and in a sense indispensable – universal and secular vision of the human" (Nigam 2020: 55). Historically, religious philosophies have played a pivotal role in shaping the habitual existential patterns of individuals and communities in India and have been co-functioning in entanglement with logic, science, technology, and modernity. Blessings are sought when new homes, office spaces, electronic devices, and digital gadgets are inaugurated, such as prayers and offerings of fruits, foods, vegetables, and flowers to different deities. Such an approach is considered to be spiritually pious and healing in nature. However, on many occasions, after the rituals are over, the offerings are disposed of into the streets, 'safely' away from the abovementioned spaces and places in locations that are not culturally, spiritually, and ideologically rich enough, like the residential areas of the sweepers, open water bodies, unclaimed lands, and others.

This interplay of imagined cultural purity and spiritual richness with the unsystematic discarding of waste objects has led to the foundation of garbo-philosophies or the practice of garbage philosophies. Regarding the arguments on different localities and religious institutions in the previous chapters, garbage philosophies evolve through rational blindness, where the paradigms of rationality and diversity are regarded as barriers to intellectual growth and development. As a result, a group of self-selected and self-celebratory endorsers of development, smartness, and modernity in contemporary India have generated a whole philosophical constellation of 'irrational rationality' where blind faiths are regarded as more rational than concrete evidence. It can be further understood with regard to the instances of triggering Islamophobia in India during COVID-19 through spreading fake news like a Muslim fruit vendor in Madhya Pradesh spitting on fruits before selling them, a group of Dawoodi Bohra Muslims spitting on their plates after finishing their lunch, and in various other ways (Chattopadhyay 2020). Further instances can be cited, like the killing of Muslims for consuming beef, killing and raping Dalits for 'contaminating' high-caste localities, mistreating low-caste students in schools, normalizing rape cultures in the name of ethics, discipline, and punishments, etc.

These instances show that wastes, as objects and ideologies, gradually percolate within humans' physical, ideological, and philosophical existence to function no differently from toxic discards – irritating, alienating, disgusting, and degrading in nature. Garbo-philosophical practices in contemporary India primarily stem from the "ideology of conservatism" of the right-wing BJP rule that prefers the maintenance of the "current social order and structure by preservation of tradition" (Keersmaecker et al. 2023: 3). In the name of constructing a traditional and unified knowledge system, BJP leadership has been carving coherent belief systems (Azevedo et al. 2019), where coherence is generated in a dictatorial and exclusionary fashion. If the ideologies and philosophies of individuals and communities fit within the culturally, communally, and politically Hinducentric vision of the BJP, they would be accepted and acknowledged; otherwise, they would be subjected to erosions, pogroms, and erasures. Such problematic attitudes are gradually pushing human civilization towards a garbocentric and garbocratic future, where every aspect of physical, intellectual, and ideological existence would be regulated and dictated by empires of waste in visible, invisible, tangible, and intangible ways. This intertwined philosophical existence of wasted humans, wasted ideologies, wasted objects, and wasted psyches is further authenticated by producing distorted and manipulated forms of garbo-histories.

Garbo-history

How about reading the history of the Taj Mahal by erasing the history of the Mughal emperor Shah Jahan? How about reading the history of the Mughal-Maratha wars by erasing the history of the Mughals? How about reading the history of anti-British resistance in the Malabar region of Kerala by erasing the history of the Mappila rebellion? How about reading about Mahatma Gandhi's assassination without knowing/naming the assassin? How about reading the history of riots in India without knowing about the 2002 riots in Gujarat, where Hindu mobs lynched several Muslims? These questions appear to be illogical, but this is what the current BJP government, as a part of its pan-Indian project of historical, intellectual, and ideological sanitization and vandalism, is doing. Any historical narrative that tarnishes and exposes the virulent sociopolitical ideologies of the BJP and glorifies the Muslims in India does not feature in the curricula and pedagogies of the schools. This is why the name of Gandhi's assassin has been removed from the history books because, politically, he was a close associate of Rashtriya Swayamsevak Sangh (RSS), the paramilitary wing of the BJP (Raj 2023). The 2002 riots in Gujarat have been removed from the history books because Narendra Modi was the Chief Minister of Gujarat (Kaur 2023). Along with these erasures, the RSS argued that untouchability was brought to India by the Muslim invaders, and the Hindus whom they enslaved were treated as untouchables (Singh 2014).

These erasures and imaginative incorporations have pushed Indian and South Asian history into a critical state of selective demolishment and disjuncture, where events are narrated without definite contextual backgrounds. The production of these baseless and illogical garbage histories is not much different from how Indian and South Asian histories were projected to the West by the European colonizers. The violent strategies of the BJP show that postcolonial institutions and structures of power sustain "colonizer-colonized relations of exploitation, domination and repression" (Ndlovu-Gatsheni 2020: 2). The exclusionary functional patterns of the BJP invade "the mental universe of the people" (Ndlovu-Gatsheni 5) and commit "crimes such as *epistemicide* (where you kill and displace pre-existing knowledge), *linguicide* (killing and displacing the languages of a people and imposing your own) and *culturecide* (where you kill or replace the cultures of a people)" (Ndlovu-Gatsheni 6). It is usually observed that when "individuals' national identity is made salient, they are less likely to acknowledge the wrongdoings that are part of their nation's history" (Eason et al. 2020: 10). When the wrongdoings are challenged, individuals get anxious about the aspect that "national representations may change" (Eason et al. 11). The production and preservation of such problematic attitudes in contemporary India have led to the formation of garbo-histories that are garbage-like, intangible, infectious, damaging, manipulative and mobile to such an extent that they often serve as a "vector of social mobility and personal success" for specific sociopolitical organizations (Highet & Del Percio 2021: 127).

These propositions and practices of fictitious sciences, philosophies, and histories have generated multiple "collateral casualties" (Z. Bauman 39) through deaths, distortions, and erasures of knowledge systems and community values. The casualties are systematized by developing research avenues like cow sciences and Hindu Studies. In 2021, the BJP-led education ministry in India proposed research areas and research centres on cow sciences to justify the "special powers" (Gettleman and Raj 2021) and "medical significance" (Mogul 2021) of cows. Later on, due to protests across India, the proposal was revoked. Apart from cow sciences, undergraduate and postgraduate courses on Hindu Studies have been launched in higher educational institutions like Banaras Hindu University, Delhi University, and others (Special Correspondent 2022; Telangana Today 2023). Under the blanket of reviving the Indigenous histories, sciences, philosophies, and cultures of India, the BJP is consistently "furthering its ideological agenda of creating Hindu Rashtra (i.e., state)" (Ramachandran 2020: 15) underpinned with garbocentric, wasteful "deeply exclusionist" and "discriminatory" visions (Ramachandran 16). These violent ideological visions are conceived as a "sacred duty" (Siyech and Narain 2018: 190), and in the name of a decolonial turn towards a neo-indigenous India,[1] they eventually manufacture enormous volumes of intellectual wastes as garbo-sciences, garbo-histories, and garbo-philosophies. However, the intentions and impact of these problematic knowledge systems reveal that instead of any decolonial turn, India is gradually undergoing a paracolonial turn. As a part of the paracolonial turn, European colonial ideologies are re-performed at the local level through internally unleashing hierarchical, exclusionary, and fractured sociopolitical practices. In a paracolonial society of wasted knowledge, the bodies of individuals and communities may

---

1   The term neo-indigenous refers to the BJP's self-inflating, self-appreciative, and self-profiting sociopolitical strategies to revive the Indigenous social, cultural, and political values of pre-colonial times through selective cultural inclusions and erasures in the name of caste, class, religion, ethics, and discipline.

belong to postcolonial spaces, but colonial spectres govern their spirits, psychologies, and ideologies.

'*Sollen*': Somewhere, Somewhat, Sometime …

When wasted knowledge systems become the general norm and garbage intellectualities are regularly celebrated, then an entire civilization of humans captivated are driven by spectres of redundancy (Z. Bauman 97) and underfed with what German philosopher Siegfried Karacauer argues as the phenomenon of "*Sollen* ('shoulds')" – the ideas and knowledge that intend "to become reality themselves" (Kracauer 1995: 143) and pretend to be so. But, as discussed in this chapter, they consistently fail in logic and reason. Today, while devising ideal governing strategies and policies in India, there is no lack of 'shoulds' as can be seen during public meetings by political parties, the release of annual budgets, and the condemnation of human and environmental catastrophes, where often-not-achievable ideal perspectives of what 'could' have been done, what 'should' have been done, and what 'needs' to be done are casually fleshed out, without contemplating the genuine possibilities.

To further contextualize this argument, when Dalit women are raped and killed by high-caste men, the reluctance of the police to accept complaints from victims' families, avoiding medical tests of raped victims, and burning of the dead bodies of the victims without informing the families (Zafar & Anand 2020; Press Trust of India 2023) are justified through a series of 'shoulds' that are embellished with the toxic narratives of ethics and discipline ('the victim should not have walked alone in the streets'); legality ('the police should have registered the complaint'); decency ('the victim should not have worn body-exposing clothes'); traditionality ('the victim should have remained inside the house and not freely roamed the streets'); and others. In this way, specific communities, ideologies, and existential patterns are normatively treated as "supernumerary, unneeded, of no use – whatever the needs and uses are that set the standard of usefulness and indispensability" (Z. Bauman 12). While weighing the dispensability and indispensability of human existential values, what goes unrealized is that in the process of reducing selective individuals and communities to wastes, the identities and existence of both perpetrators and victims become like discarded objects, a symbolically stinking, ideologically overhauling, and physiologically nauseating heap of worthlessness.

The physical, ideological and intellectual beings should not be restricted within the fixated compartments of popular ideologies but should be regularly subjected to "moulding, kneading, squeezing and stretching" (Z. Bauman 19). However, "the roots of the trouble, it seems, have moved further away than we can reach" (Z. Bauman 17), eventually forming garbo-power and garbo-ideologies in India that are institutionalized, metaphorized, and injected through toxic leaderships.

Works Cited

ANI. (2022). "India Is Restoring Its Glory and Prosperity, Entire World Will Benefit: PM Modi", *The Print*, <https://theprint.in/india/india-is-restoring-its-glory-and-prosperity-entire-world-will-benefit-pm-modi/1163928/>, accessed on 21 June 2023.

Azevedo, F. et al. (2019). "Neoliberal Ideology and the Justification of Inequality in Capitalist Societies: Why Social and Economic Dimensions of Ideology Are Intertwined", *Journal of Social Issues*, 75, 49–88.

Bakhtin, M. (1968). *Rabelais and His World*. Massachusetts: MIT Press.

Bauman, Z. (2004). *Wasted Lives: Modernity and Its Outcasts*. Oxford, New York & Boston: Polity Press.

Chattopadhyay, A. (2020). "Fact Check: Fake Videos Shared to Defame Muslim Community for Spreading Coronavirus", *The Logical Indian*, <https://thelogicalindian.com/fact-check/muslim-lick-plates-coronavirus-covid-19-pandemic-20429>, accessed on 31 December 2022.

Doron, A. & Jeffrey, R. (2019). *Waste of a Nation: Garbage and Growth in India*. Cambridge, Massachusetts and London: Harvard University Press.

Eason, A. E. et al. (2020). "Sanitizing History: National Identification, Negative Stereotypes, and Supporting for Eliminating Columbus Day and Adopting Indigenous Peoples Day", *Cultural Diversity and Ethnic Minority Psychology*, 27 (1), 1–17.

Gettleman, J. & Raj, S. (2021). "Do India's Cows Have Special Powers? Government Curriculum Is Ridiculed", *The New York Times*, <https://www.nytimes.com/2021/02/22/world/asia/india-cow-exam-curriculum.html>, accessed on 28 September 2022.

Gray-Cosgrove, C., Liboiron, M. & Lepawsky, J. (2015). "The Challenges of Temporality to Depollution & Remediation", *S.A.P.I.EN.S. Surveys and Perspectives Integrating Environment and Society*, 8 (1), <https://sapiens.revues.org/1740>, accessed on 21 February 2021.

Highet, K. & Del Percio, A. (2021). "When linguistic capital isn't enough: Personality development and English speakerhood as capital in India". In J. E. Petrovic & B. Yazan (eds.), *The Commodification of Language: Conceptual Concerns and Empirical Manifestations*, pp. 127–143. London & New York: Routledge.

Kaur, A. (2023). "New Indian Textbooks Purged of Nation's Muslim History", *The Washington Post*, <https://www.washingtonpost.com/world/2023/04/06/india-textbooks-muslim-history-changes/>, accessed on 4 October 2023.

Keersmaecker, J. D. et al. (2023). "Rationally Blind? Rationality Polarizes Policy Support for Colour Blindness versus Multiculturalism", *British Journal of Social Psychology*, 1–17.

Kracauer, S. (1995). *The Mass Ornament: Weimar Essays*. Edited and Translated by T. Y. Levin. Harvard: Harvard University Press.

Lepawsky, J. (2018). *Reassembling Rubbish: Worlding Electronic Waste*. Massachusetts: MIT Press.

Lillie, W. (1957). *An Introduction to Ethics (3rd Edn)*. London: Methuen & Co Ltd.

MacBride, S. (2012). *Recycling Reconsidered: The Present Failure and the Future Promise of Environmental Action in the United States*. Cambridge: MIT Press.

Mogul, R. (2021). "Cows Are Sacred in India. Critics Say a New National Exam Politicizes the Animal", *CNN*, <https://edition.cnn.com/2021/01/08/india/cow-science-exam-india-intl-hnk-scli/index.html>, accessed on 14 August 2023.

Narayanan, N. (2014). "What Narendra Modi Has to Say About Climate Change." *Quartz*, <https://qz.com/india/269743/has-pm-modi-changed-his-position-on-climate-change-from-when-he-was-gujarat-cm>, accessed on 25 December 2018.

Naveen. (2023). "Garbage and Vulnerability", Personal Conversation, Belgachia.

Ndlovu-Gatsheni, S. J. (2020). "Decolonization, Decoloniality, and the Future of African Studies: A Conversation with Dr. Sabelo J. Ndlovu-Gatsheni – Interviewed by Duncan Omanga", *Items: Insights from the Social Sciences*, <https://items.ssrc.org/from-our-programs/decolonization-decoloniality-and-the-future-of-african-studies-a-conversation-with-dr-sabelo-ndlovu-gatsheni/>, accessed on 31 March 2023.

Nigam, A. (2020). *Decolonizing Theory: Thinking Across Traditions.* New Delhi: Bloomsbury.

Press Trust of India. (2023). "Dalit Woman Raped, Set on Fire in Rajasthan's Barmer, Dies of Burn Injuries", *NDTV*, <https://www.ndtv.com/india-news/dalit-woman-raped-set-on-fire-in-rajasthans-barmer-3930925>, accessed on 30 October 2023.

Press Trust of India. (2023). "'Golden Period' of PM Narendra Modi Will Be Taught to Future Generations: Mansukh Mandaviya", *Outlook*, <https://www.outlookindia.com/national/-golden-period-of-pm-narendra-modi-will-be-taught-to-future-generations-mansukh-mandaviya-news-263433>, 6 January 2024.

Rahman, M. (2014). "Indian Prime Minister Claims Genetic Science Existed in Ancient Times." *The Guardian*, <https://www.theguardian.com/world/2014/oct/28/indian-prime-minister-genetic-science-existed-ancient-times>, accessed on 27 March 2021.

Raj, S. (2023). "New Indian Textbooks Purged of Muslim History and Hindu Extremism", *The New York Times*, <https://www.nytimes.com/2023/04/06/world/asia/india-textbooks-changes.html#:~:text=Chapters%20on%20Mughal%20history%2C%20covering,several%20attempts%20to%20assassinate%E2%80%9D%20him>, accessed on 31 December 2023.

Ramachandran, S. (2020). "Hindutva Violence in India: Trends and Implications", *Counter Terrorist Trends and Analysis*, 12 (4), 15–20.

Scroll Staff. (2015). "Holy Cow! Recent Bovine News You May Have Missed", *Scroll.in*, <https://scroll.in/article/717420/holy-cow-recent-bovine-news-you-may-have-missed>, accessed on 20 January 2023.

Scroll Staff. (2018). "Darwin Is Scientifically Wrong, Says Minister Satyapal Singh, BJP Leader Ram Madhav Supports Him", *Scroll.in*, <https://scroll.in/latest/865803/nobody-saw-ape-turning-into-man-charles-darwins-theory-of-evolution-is-wrong-union-minister>, accessed on 21 March 2022.

Singh, D. K. (2014). "RSS Rewrites History: Dalits 'Created' by Invaders", *Hindustan Times*, <https://www.hindustantimes.com/india/rss-rewrites-history-dalits-created-by-invaders/story-eyBt99Y2XbICUbadzCsSkM.html>, accessed on 16 April 2022.

Siyech, M. S. & Narain, A. (2018). "Beef-related Violence in India: An Expression of Islamophobia", *Islamophobia Studies Journal*, 4 (2), 181–194.

Special Correspondent. (2022). "BJP MP Moves Bill to Revive 'Golden Age of Ancient India'", *The Hindu*, <https://www.thehindu.com/news/national/bjp-mp-moves-bill-to-revive-golden-age-of-ancient-india/article65259601.ece>, 2 January 2024.

Special Correspondent. (2022). "Hindu Studies: Third Central University Launches Master's Programme", *The Telegraph*, <https://www.telegraphindia.com/india/hindu-studies-third-central-university-launches-masters-programme/cid/1890259>, accessed on 19 July 2023.

Telangana Today. (2023). "DU to Launch a Course on Hindu Studies", *Telangana Today*, <https://telanganatoday.com/du-to-launch-a-course-on-hindu-studies>, accessed on 17 August 2023.

Teo, M. M. M. & Loosemore, M. (2010). "A Theory of Waste Behaviour in the Construction Industry", *Construction Management and Economics*, 19 (7), 741–751.

Vyas, B. P. (2017). "Cows Inhale and Exhale Oxygen, Says Rajasthan Minister; Twitter Explodes", *NDTV*, <https://www.ndtv.com/india-news/cows-inhale-and-exhale-oxygen-says-rajasthan-minister-twitter-explodes-1649920>, accessed on 30 November 2022.

Wacquant, L. (2001). "Comment la 'tolerance zero' vint a l'Europe", *Maniere de Voir*, 37–51.

Zaffar, H & Anand, K. (2020). "Indian Dalit Rape Victim Family Locked Up as Police Burned Body". *Al Jazeera*, <https://www.aljazeera.com/news/2020/10/2/hathras-they-locked-us-inside-our-home-and-burnt-her-body>, accessed on 24 June 2023.

CHAPTER 4

Garbo-power and Garbo-being: The Onset of Garbo-ideologies

> "However artificially the Partition may have been made, it has to be accepted."
> (Bettelheim 1968, xv)

Feku-Fakeness-Fetishism: The Toxic Trident

In the book *India Independent*, Charles Bettelheim said the abovementioned words in the context of the India-Pakistan partition. These words continue to be significant within the socially, culturally, politically, communally, and religiously fractured regime of the right-wing government of the BJP in India. In 2023, India organized the G20 Summit, themed as 'One Earth, One Family, One Future.' During the summit, Narendra Modi justified the theme's relevance by saying that "India led the renewable energy revolution with One Sun, One World, One Grid. India strengthened the global health initiative with One Earth, One Health. And now India's theme for the G20 will be One Earth, One Family, One Future" (Modi quoted in India Today Web Desk 2022). However, under the BJP-led government, the rapid degradation of the natural environment due to reckless industrialization, violation of environmental laws, and division of communities based on castes and religions draw a contradictory picture.

In 2014, the then environment minister of India, Veerappa Moily, gave a "green nod" to the South Korean steel joint POSCO's $12 billion steel plant projects in Orissa, which faced long-standing and stiff on-the-ground opposition besides legal tussles related to environmental approvals

(Brinda 2017). In the same year, the "number of independent members in the National Board of Wildlife (NBWL) was reduced from 15 to just three" (Rathee 2019). The new NBWL was a mere puppet in the hands of the government. As a result, in the next five years, "NBWL approved 99.82 percent of all industrial projects, giving them environmental clearance" (Rathee 2019). It legitimized the processes of environmental destruction by "kowtowing to the government's agenda of development at all costs" (Bindra 2017: xv). In 2017, the "Central Pollution Control Board wrote to over 400 thermal power units in the country, allowing them to release pollutants in violation of the 2015 limits set by the government, which were to be followed for another five years" (Rathee 2019). By 2018, 15 of the 20 most polluted cities in the world were found in India (Koshy 2019). In 2021, the BJP government changed the forest conservation laws to construct highways and railroads through the forest. Consequently, India has witnessed "enormous forest fires and 20% forest degradation in the past years" (Pal 2021).

These circumstantial evidences must provoke us to question the thematic validity and intention of the summit. What is more surprising is the acceptance, celebration, idealization, and idolization of this fake thematic vision by the G20 members. The global "distribution, dissemination, and drift" (Ghosh 2022: 10) of the G20 principles took place with an "algorithmic perspective" (Van Herdeen & Bas 2021: 177). They were conceived and laid out in an "(in)calculable, differential, equational, and relational" manner (Ghosh 168). All the G20 member countries are united by a common challenge of countering the resistance of Indigenous communities and cultures against disruption and erasures of the natural environment and resources. It is also known how the draconian and garbage governing policies in India and other member countries, in the name of procuring "equilibrated configurational freedom" (Ghosh 57), have been generating indexes of "complex forces and technologies" (Buck 39), underlined with the toxic trident of *'feku*-fakeness-fetishism.' The term *feku* has explicitly been used to denote the political leaders of the BJP government who have been making fake promises and manufacturing fake data to justify their promises. The fakeness of the BJP government has been applauded and fetishized by not only the right-wing neoliberal socioeconomic organizations in India but also by the organizations in the Global North, as the latter could find appropriate allies to systemically continue with their neocolonial and paracolonial modes of knowledge production and confirm the "ending of all hope" (Drudi 1972: 36). The trident has been lubricated by a "rubbish ecology" (Yaeger 2008) of wasted environments, social dynamics, power structures, beings, and ideologies, which has been turning "habits, lifestyles, emotions, and economy" (Ghosh 13) in India towards a disgusting and detrimental mould of garbage mentalities and governmentalities. Like discarded objects, the trident exists in a state of "inherent formlessness" of "seamless curvature" and "free-flowing undulation" (Buck 35). It is a living entity, an invasive structure, an invisible dictator, an imposter, and lives simultaneously through the past, present, and future.

The tendency of the trident to deliberately neglect genuine, open-minded, and critical education and research in contemporary times and restrict knowledge-making facilities within a limited group of socioeconomically privileged classes of people can be traced back to the neglect of the British administration in the colonial era. Their aim was "never to raise the educational standard of the whole population" but to "fill the subordinate administrative posts essential for the continuance of 'good' colonial administration as well as the lower ranks of what were considered to be essential industries" (Bettelheim 16). During the colonial era, "a great deal of room was left for private education" in the cities, and it "often tended to be costly" (Bettelheim 16).

On a similar note, in the name of globalizing and internationalizing the higher education system of India, the exclusionary and censored functional procedures of the high-priced, elitist, neoliberal, pro-right-wing, and socio-politically gatekept institutions are regarded as the ideal parameters of intellectual progress. For instance, let us look into how Ashoka University has been subjected to funding restrictions and intellectual censorship for being critical of the BJP government. Sabyasachi Das, a faculty at Ashoka University, was forced to resign from his post in August 2023 for exposing how, under the BJP regime, "manipulation appears to take the form of targeted electoral discrimination against India's largest minority group – Muslims, partly facilitated by weak monitoring by election observers" in a research paper (Das 2023). In September 2023, another faculty from the same university named Gilles Verniers was forced to leave the university for not conforming with the right-wing neoliberal functional turn of the Trivedi Centre for Political Data, of which he was the founder and co-director (Scroll Staff 2023). These instances show that garbage ideologies function as one of the foundations of the governing procedures of the central government in India, and they are "visible, haptic, diagrammatic or geometric, referential, structural, curvilinear, differential, and hence, figural" (Ghosh 5). The ideologies are "truly synthetic and tailored to meet specific needs" (Ghosh 13; also see Walker 1994: 67). To further delve into the evolutionary and functional patterns of the garbo-ideologies, it is vital to investigate the materiality and the immateriality of such ideologies.

Material Garbo-ideologies

Material Garbo-ideologies are "developmental, mutative, mimetic, adaptive and transgenic" (Ghosh 40). They are portrayed and proliferated through posters, paintings, t-shirts, nomenclatures, architecture, textbooks, policy books, flags, information boards, etc. These materials as garbage epitomize "the ephemeral, the ever changing" (Bensaude-Vincent 2013: 23). During the pre-election campaigns in 2013 and 2014, the BJP used diverse physical and digital tactics to present themselves as a perfect solution to India's post-independent battle against the Euro-North American social, cultural, political, economic and industrial encroachments. But, quite ironically, the BJP government borrowed campaigning techniques from the West. For instance, during my stay in Varanasi, I observed that, besides distributing flyers about the evolution and success of the BJP party and conversing with people who are socio-politically identified as neglected and marginalized, like the sweepers and ragpickers, the party workers distributed t-shirts, policy books, flags, and miniature paintings of Hindu gods and goddesses to the passersby.

After a certain period, these objects were commonly sold in the shops for political and ideological promotions on the one side and for commercial benefits on the other. Many shopkeepers shared that they were forced to hand over a part of the profit they made by selling different objects associated with the signs, symbols, and ideologies of the BJP party to the local BJP goons as tax. If they refused to pay the tax, they were threatened with eviction. Besides paying the tax, many were forced to vote for the BJP party during the elections. This association of politics, commerce, and unethical extractions during election campaigns was also prevalent during the British colonial era in India when the high-caste and high-class communities extracted 'welfare' taxes from the low-class and low-caste communities (Bhambra 2022).

The political propaganda of the BJP was also digitally marketed by preparing music videos on the pseudo-success history of the party, making documentaries about the monopolistic sociopolitical prowess and capability of a Hinducentric India, using AI applications to portray BJP leaders as godlike Hindu spiritual and mythological characters; curating fake videos of the sufferings of Hindus at the hands of Muslims, and in many other ways (Thaker 2018; Chaudhari 2020). Like the physical media, the pre-election campaigns through digital media have also proven to be socio-politically and commercially successful for the BJP.

Apart from implementing these colonially mimicked strategies, the BJP government, like the colonizers, has also selectively erased the Muslim names of various railway stations, streets, educational institutions, and work organizations across the country as part of their pan-Indian project of traditionalizing and re-indigenizing India. In these impositions, transgressions, and transformations, things like t-shirts, miniature paintings, policy books, and others no longer function as mere objects. Instead, they are transfigured into violent garbocratic weaponries of "debased, deceased, discarded, decrepit and downgraded line of objects" (Ghosh 86).

Generally, garbage as physical entities can "reveal two types of cultures: those that squander resources on material symbols of conspicuous consumption, and those that stretch out the use-life of resources" (Hansen 2020: 25). These cultures, by wielding diseased suffocating, infectious, bacterial, and wasteful patterns of human existence in India, have entailed a "destructive negation" of selective physiologies, ideologies, and lifestyles (Halland 2019: 43). Garbage ideologies in their material form "go beyond the attractions of the imaginations of forms; [it] thinks matter, dreams in it, lives in it, in other words materializes the imaginary" (Bachelard quoted in Harris 2018: 140). They provide "particular ways of thinking about and advancing understandings of materiality *as process*" (Gabrys, Hawkins & Michaels 2013: 3). The material garbo-ideologies through various social, cultural, political, and economic agencies are being preserved through "relational mattering," underpinned by "provisionality, permanence, plurality, and the precarity" of wasteful thoughts. They entrench themselves so profoundly and normatively within human bodies and psyches that in the process of resisting deformations, humans consciously and unconsciously yield to deformations of their existence (Malabou 2005: 9). This simultaneous process of receiving and giving form to garbo-powers, garbo-beings, and garbo-ideologies takes place not only materially but immaterially as well.

Immaterial Garbo-ideologies

Immaterial garbo-ideologies are "material-philosophical-aesthetic" (Ghosh 9). They are tangibly intangible and intangibly tangible. They generate illusionary feelings of being touched and felt but can only be emotionally, symbolically, and semiotically perceived, internalized, and interpreted. Immaterial forms of garbo-ideologies are polymorphous and malleable. They are characterized by subtle " 'suppleness' and flexibility" (Malabou 8), as can be seen in the case of the silences of the BJP ministers when Dalit women are raped, Muslims are publicly lynched, mosques are vandalized, and nature and wildlife are destroyed on the one side, and the justification of these acts through the narratives of development, ethics, morality, respect and traditionality on the other. Before the 2019 elections, the Prime Minister of India launched 'Gram Swaraj' (self-governed villages) programmes and directed his cabinet leaders to spend time with socioeconomically marginalized communities like Dalits in the villages.

Many BJP leaders went to spend time and dine with Dalit families. Newspapers were flooded with images of BJP leaders and Dalit people sitting and smiling together, thus projecting a new India of inclusion, mutual respect, and togetherness. However, later on, investigations from certain media houses revealed that the images of sitting, dining, and smiling together were fake as food and mineral water were ordered from outside to consume at Dalit homes, and the Dalits were made to 'specially' clean their houses so that the high-caste BJP leaders did not get spiritually defiled by their presence (Anshuman 2018; Kaushal 2018). When the leaders were questioned about these divisive, artificial, wasted, and fraudulent practices of togetherness, they remained silent, neither accepting nor defending their persistent fetishism towards fakeness. Much later, a few leaders feigned ignorance and plastically apologized. As a part of the immaterial sociopolitical garbo-ideologies of the BJP, silence and ignorance serve as essential tools of propagandizing. Ignorance is not an innocent and casual act but is "actively constituted and reproduced as an act of power" (Feenan 2007: 511). It is a medium "for the production and maintenance of unequal positionalities in a society, with the result that, like knowledge, its distribution can be mapped along societal fault lines" (Steyn 2012: 10).

Concerning the propagation of immaterial garbo-ideologies, another significant example could be the 'Mann Ki Baat' programme of Prime Minister Narendra Modi. The programme was initiated in 2014 as a monthly radio programme to publicly discuss and address pertinent issues that the nation is commonly facing. However, with time, it became another medium for the BJP to preach and promote its Hindumaniac and Islamophobic propaganda by presenting distorted political and scientific facts and strategically ignoring the regular violence against women, Muslims, and Dalits. To build fast and fake popularity of the programme across the country, the local BJP leadership compelled the schools to make the students and teachers listen to the 'Mann Ki Baat' programme. Those institutions that did not abide by the mandate were given show-cause notices and threatened with punishments and loss of affiliations (TNN 2023; Mohanty 2023). Like publicly discarded wastes, irrespective of the irrelevance, toxicity, disintegrity, plasticity, and vengeful attitude of the programme, forceful "junk bonds" (Ty 2015: 628) of fake ideologies, expectations, assurances, and outcomes have been generated between the BJP government and the citizens, which ultimately has led to the formation of wasteful, molecular, synthetic, polymerized, and ephemeral garboglomerates.

Garboglomerates

Garbage as symbolical objects and objective symbols has become a "figure of the universal" (Chakraborty 2009: 222) – a "creature in the becoming" and growing out of imminent catastrophes (Ghosh 37). Garbo-ideologies as mass democratic ideologies in India have emerged successfully by aiming at "something common, not rare" (Barthes 1993: 117). It generates its own "emergence, usability, denouement, non-utility, and materialization through an undulating trajectory of manifold becomings, ecological materiality, and eco-cultural occurrence" (Ghosh 111). The functionalities of material and immaterial garbo-ideologies have led to the evolution of garboglomerates or conglomerates of humans, who physically, ideologically, visibly, and invisibly function as garbage by continuously "weaving, netting, knitting, and braiding" piles of discarded thoughts, mindsets, psyches, powers and beings (Buck 2011: 40). Garboglomerates can "intersect, entangle, constellate and trajectorize" (Ghosh 8) and the dangers of garboglomerates lie in their character of "morphing, adjusting, experimenting and moulding itself to fit cut shapes and spaces." (Ghosh 18). Like infectious bacteria, they multiply through binary fission – growing, breaking, reconfiguring, and expanding. Their ability to "incorporate, constellate, and aggregate" and exhibit resistance to "perishing and ruin" make them both present and absent at the same time (Ghosh 22).

'Rhizomatic Garbo-entanglements'. Artwork by Sayan Dey.

'Choked!!! A Garboglomerate'. Artwork by Sayan Dey.

To understand further, as discussed in this chapter, garboglomerates very subtly, systemically, and convincingly construct physical and symbolical vestibules of dictatorial power structures within societies, families, educational institutions, and workplaces, where garbo-powers and garbo-beings operate in an invisible and hauntological pattern, "both *leaking* of itself and *leaking* into something else" (Ghosh 25). The individuals who are part of garboglomerates may think, talk, and walk around like human beings, but like wastes, they function in "synthetic, artificial, assimilative, aggregative, and co-constructive" ways (Ghosh 26). To be a part of garboglomerate, one must consciously surrender one's capacities of liberal, critical, and multiversal thinking and "live in a world of others' words" (Bakhtin 1986: 143; also see R. Bauman 2004). This has made garboglomerates "unreliable, inconsistent, and unstable" (Ghosh 2) and they have been nurturing a "highly polarizing growth process" (Mazumdar 2017: 120) in present-day India in terms of reckless privatization, environmental destruction, violent communalization, and erasures of selective knowledge systems. Besides garbo-powers, garbo-beings, and garboglomerates, the social, cultural, political, and economic polymerization and polarization in India are further triggered by garbage pedagogies and garbage curricula in the schools and higher educational institutions, which have been discussed in the following chapter.

Works Cited

Anshuman, K. (2018). "'Eating with Dalits' Is Only a Photo-op and Election Strategy for the BJP", *The Print*, <https://theprint.in/politics/how-eating-with-dalits-became-a-farce-and-why-the-bjp-should-be-worried/54729/>, accessed on 13 March 2024.
Bakhtin. M. M. (1986). *Speech, Genres, and Other Late Essays*. Translated by V. W. McGee. Austin: University of Texas Press.
Barthes, R. (1993). *Mythologies*. Translated by A. Lavers. London: Vintage.
Bauman, R. (2004). *A World of Others' Words: Cross-Cultural Perspectives on Intertextuality*. Oxford: Blackwell Publishing.
Bensaude-Vincent, B. (2013). "Plastics, Materials and Dreams of Dematerialization", In J. Gabrys, G. Hawkins, & M. Michael (eds.), *Accumulation: The Material Politics of Plastic*, pp. 17–29. New York & London: Routledge.
Bettelheim, C. (1968). *India Independent*. New York: Monthly Review Press.

Bhambra, G. K. (2022). "Relations of Extraction, Relations of Redistribution: Empire, Nation and the Construction of the British Welfare State", *British Journal of Sociology*, 73 (1), 4–15.

Bindra, P. S. (2017). *The Vanishing: India's Wildlife Crisis*. Gurugram: Penguin Viking.

Buck, B. (2011). "What Plastic Wants." *Log*, 23, 35–40.

Chakraborty, D. (2009). "The Climate of History; Four Theses", *Critical Inquiry*, 35 (2), 197–222.

Chaudhuri, P. (2020). "Amit Malviya's fake News Fountain: 16 Pieces of Misinformation Spread by the BJP IT cell Chief", *Scroll.in*, <https://scroll.in/article/952731/amit-malviyas-fake-news-fountain-16-pieces-of-misinformation-spread-by-the-bjp-it-cell-chief>, accessed on 29 April 2021.

Das, S. (2023). "Democratic Backsliding in the World's Largest Democracy", *SSRN*, <https://papers.ssrn.com/sol3/papers.cfm?abstract_id=4512936>, accessed on 25 July 2023.

Drudi, G. (1972). "Design and Landscape in Italy", *Craft Horizons*, 32 (4), 31–42.

Feenan, D. (2007). "Understanding Disadvantage Partly through an Epistemology of Ignorance", *Social & Legal Studies*, 16 (4), 509–531.

Gabrys, J., Hawkins, G. & Michael, M. (2013). "Introduction: From Materiality to Plasticity". In J. Gabrys, G. Hawkins, & M. Michael (eds.), *Accumulation: The Material Politics of Plastic*, pp. 1–14. New York & London: Routledge.

Ghosh, R. (2022). *The Plastic Turn*. Ithaca: Cornell University Press.

Halland, I. (2019). "Being Plastic", *Log*, 47, 35–44.

Hansen, J. (2020). "An Abundance of Fruit Trees: A Garbology of the Artifacts in *Animal Crossing: New Leaf*, Loading …", *The Journal of the Canadian Games Studies Association*, 13 (22), 23–38.

Harris, P. A. (2018). "Stoned Thinking: The Petriverse of Pierre Jardin", *Substance*, 47 (2), 119–148.

India Today Web Task. (2022). "'One Earth, One Family, One Future': PM Modi unveils India's G20 mantra", *India Today*, <https://www.indiatoday.in/india/story/pm-modi-unveils-india-g20-mantra-logo-theme-website-2294839-2022-11-08>, 8 November 2022.

Kaushal, R. (2018). "What Made BJP Think Dining at Dalit Houses Was Panacea for the Community's Problems?", *Newsclick*, <https://www.newsclick.in/what-made-bjp-think-dining-dalit-houses-was-panacea-communitys-problems>, accessed on 25 July 2023.

Koshy, J. (2019). "Fifteen of the 20 Most Polluted Cities in the World Are in India", *The Hindu*, <https://www.thehindu.com/sci-tech/energy-and-environment/fifteen-of-the-20-most-polluted-cities-in-the-world-are-in-india/article26440603.ece>, accessed on 13 March 2024.

Malabou, C. (2005). *The Future of Hegel: Plasticity, Temporality and Dialectic*. Translated by L. During. New York & London: Routledge.

Mazumdar, S. (2017). "Neo-Liberalism and the Rise of Right-Wing Conservatism in India", *Desenvolvimento em Debate*, 5 (1), 115–131.

Mohanty, B. K. (2023). "36 Nursing Students Punished for Not Listening to Prime Minister Modi's Mann Ki Baat", *The Telegraph India*, <https://www.telegraphindia.com/india/36-nursing-students-punished-for-not-listening-to-prime-minister-modis-mann-ki-baat/cid/1936231>, Accessed on 12 May 2023.

Pal, S. (2021). "World Environment Day: How is Modi Govt Faring?", *Newsclick*, <https://www.newsclick.in/world-environment-day-how-modi-govt-faring>, 05 June 2023.

Rathee, D. (2019). "Environment is the Most Under-Reported Disaster of Narendra Modi Government." *The Print*, <https://theprint.in/opinion/environment-is-the-most-under-reported-failure-of-narendra-modi-government/223670/>, accessed on 19 April 2023.

Scroll Staff. (2023). "Gilles Verniers 'forced to leave' Ashoka University's Trivedi Centre for Political Data, Says Board", *Scroll.in*, <https://scroll.in/latest/1055841/gilles-verniers-forced-to-leave-ashoka-universitys-trivedi-centre-for-political-data-says-board>, accessed on 12 September 2023.

Steyn, M. (2012). "The Ignorance Contract: Recollections of Apartheid Childhoods and the Construction of Epistemologies of Ignorance", *Identities: Global Studies in Culture and Power*, 19 (1), 8–25.

Thaker, A. (2018). "In India, BJP Supporters Are More Likely than Others to Share Fake News." *Quartz*, <https://qz.com/india/1461262/indias-bjp-supporters-share-more-fake-news-than-others-says-bbc>, accessed on 13 November 2018.

TNN. (2023). "Dehradun: School Asks Rs 100 from Students Who Missed Mann Ki Baat." *The Times of India*, <https://timesofindia.indiatimes.com/city/dehradun/dehradun-school-asks-rs-100-from-students-who-missed-mann-ki-baat/articleshow/100025433.cms?from=mdr>, accessed on 03 May 2023.

Ty, M. (2015). "Trash and the Ends of Infrastructure", *Modern Fiction Studies*, 61 (4), 606–630.

Van Herdeen, I & Bas, A. (2021). "AI as Author Bridging the Gap between Machine Learning and Literary Theory", *Journal of Artificial Intelligence Research*, 71, 175–189.

Walker, A. (1994). "Plastics: The Building Blocks of the Twentieth Century", *Construction History*, 106, 7–88.

Yaeger, P. (2008). "The Death of Nature and the Apotheosis of Trash; Or Rubbish Ecology", *PMLA*, 123 (2), 321–339.

SECTION III
Distributing

CHAPTER 5

Garbo-pedagogies and Garbo-curricula: Schools as Knowledge-garbage Laboratories

> "… waste symbolizes the inherent corruption upon which every society is built."
> (Iovino 2009: 340)

"Vyaneeshh!": Schools as Concentration Camps, Gas Chambers, and Mumbo-jumbo Rooms

"Vyaneeshh!"[1] – a famous line by the magician in Satyajit Ray's politically satirical film *Hirok Rajar Deshey* (1980) – became a word of terror for farmers, labourers and teachers in 'Hirak Rajya' (the kingdom of diamonds) because whenever they resisted the king's dictatorship and censorship, they were captured, imprisoned in the mumbo-jumbo room, and hypnotized as enslaved people. As a result, they lost the capacity to think and act individually. The educational and intellectual conditions in contemporary India are no different. Schools in India are not much different from concentration camps, gas chambers, and mumbo-jumbo rooms.

In 2023, during a conversation with a friend of mine, who was then employed at a higher education institution in Maihar City of Madhya Pradesh, he lamented how the National Education Policy 2020 (NEP 2020) was implemented in his college and other colleges in Madhya Pradesh in a flawed and fraudulent manner. As a faculty coordinator, he travelled to various colleges in Maihar and other parts of Madhya Pradesh, observing that NEP 2020 meetings with the faculty members

---

[1] The English word 'vanish' in Bengali is often pronounced as 'vyaneeshh.'

and students were regulated and presided over by local BJP political leaders instead of academic and research experts. These leaders used academic spaces to push sociopolitical propaganda, distracting people from genuine goals of praxis-based curricular and pedagogical transformation. This occurs not only in Madhya Pradesh, but also across India, where individuals and institutions remain confused about the meaning, intentions, and methodologies behind implementing education policies. As discussed in previous chapters, the social, cultural, religious, and ideological eyewash tactics of the BJP government in India are deployed in educational institutions through policies like NEP 2020.

In the name of liberty, flexibility, internationalization, and "revamping" (MHRD 2020: 3) the higher education system in India, the BJP government has been systematizing "a practice of elimination that targets the oppositional press, revolutionary ideas, perceived enemies, migrants … troubling knowledge and historical memories that threaten the existing racist and capitalist order" (Giroux 2023). The NEP has been built upon blind mimicry of Euro-North American education systems, in which the humanities and social sciences are systematically undermined, with most financial and infrastructural investment directed toward basic sciences, technologies, business, and management disciplines. Such hierarchical and exclusionary practices aim to discard critical thinking capacities and procedures as waste and create bands and brands of physiologically distorted, narcissistic, and sterilized communities, who, through their knowledge, can only give birth to "exhaustion and irrelevance" (Ndlovu-Gatsheni 2015: 195). One goal of the NEP is to make the learning experience "holistic, integrated, enjoyable, and engaging" (MHRD 2020: 11) in schools and higher educational institutions. However, this vision is pursued through principles of "corporatisation" and "neoliberal agenda" (Mpofu and Ndlovu-Gatsheni 2020: 3), which force educational institutions into a model of "rigid leadership managerialism" (Mpofu and Ndlovu-Gatsheni 2020: 3; also see Ndlovu-Gatsheni 2013), compelling "academics to pursue short-term goals with no connection to the public interest in their teaching" (Ogachi 2011: 44).

Let us unpack these challenges one by one. As part of the holistic teaching–learning approach, the NEP instructs to "reduce curriculum content to enhance essential learning and critical thinking" through "inquiry-based, discovery-based, discussion-based, and analysis-based learning" (MHRD 12). However, these goals are pursued by erasing selective Islamic civilizational histories from school textbooks, normalizing unproven mythological scientific narratives as Indigenous sciences, labelling tribal histories and cultures as unmodern, racializing curricula and pedagogy by erasing histories and cultures of Dalit communities and communities from northeastern India, and limiting inquiry, discovery, and discussion to certain "garbocratic" individuals and institutions. These individuals and institutions choose to compromise with truth and celebrate falsehoods, constructing abusive, pervasive, wasted, illogical, and manipulative knowledge systems. For instance, in 2022, during a public address, Narendra Modi stereotypically referred to the tribal communities in India as "vanvasis" (PTI 2023), meaning 'forest dwellers,' which is a very narrow way of representing the Indigenous communities because "they are the true owners of 'jal,' jungle and 'zameen' (water, jungle, and soil). The people who use such terms (vanvasi) show their ignorance towards adivasis as well as their efforts in preserving the jungles in this country" (Pawar quoted in PTI 2023).

Dalit and northeast Indian cultures are also subjected to identical forms of stereotyping regularly. Even after 76 years of India's judicial independence from European colonization, the history textbooks in schools and higher educational institutions in India have failed to be inclusive enough to include the histories of northeast Indian communities and Dalits (Kalia 2021; Kikon 2021). These exercises of ignoring and stereotyping histories, cultures and knowledge convey a "generalised, fuzzy and homogenised understanding of the other" (Nawani 2014: 20).

The NEP also stresses the necessity of "experiential learning" through "storytelling-based pedagogy" (MHRD 12). But, in reality, stories are told in socio-historically selective, culturally erased, and psychologically derogatory patterns so that the seeds of disgust and hatred are sown within the psyches of individuals right from their childhood. What stories can be made from erased and distorted histories and cultures? What stories can be narrated through the tongues of exclusion, insult, and hatred? What knowledge can be shaped through unproven scientific narratives? These questions unveil how India's problematic educational policies have contributed to the "emergence of a new 'crisis of quality' engineered from within the institutions" (Ogachi 44). These problematic policies are gradually converting educational institutions into concentration camps, gas chambers, mumbo-jumbo rooms, and laboratories, where exams, syllabuses, teaching, and learning function as mere excuses to metamorphose human minds into propaganda machines. As part of this metamorphosis, the learners "struggle against the ravages of the consultancy syndrome that rewards reports over refereed academic papers, against the repressive practices and criminal negligence" of the national government and pressures for the "commercialisation of education systems ... " (Mkandawire 2011: 33). Eventually, individuals and institutions are subjected to a process of "cleaning up" (Benowitz 2009: 90), during which every possibility of arguing, disagreeing, and critiquing is shut down and are replaced by discarded and dictatorial mono-epistemological approaches (Mignolo 2007: 159).

The violence-based curricular and pedagogical approaches function as a "bewildering disease" and, like an infectious virus or bacterium, threaten to "reach epidemic proportions" (Bainbridge quoted in Benowitz 89). The school institutions and political organizations further maintain the concentration camps and laboratories of intellectual violence through the assistance of investment organizations which, in the name of philanthropy, fund allocation, and scholarships, decide which knowledge disciplines are worthy of support and which are not; divide learners and their career choices based on socioeconomic status rather than intellectual capacity; conduct university entrance examinations by "filtering students based on their ability to pay tuition and fees" (Boossabong 2018: 113) instead of competitive scores; and shut down selective academic departments as they are not profit-worthy. Whosoever abides by these parameters is allowed to be part of the neoliberal educational institutions, and the rest are rejected, dumped, and flushed out. As a result, to fulfil the desires of the neoliberal education system, many individuals must bear the burden of heavy loans, ultimately creating a ceaseless generation of debtors, bonded labourers, and waste humans, who are "incorporated into the new social body" that resembles a "dumping site" (Bauman 2004: 77). There is no scope of return or "no road forward" from that site (Bauman 77). Those who try to escape from the site and its neoliberal entrapments are pushed into a lifelong state of "hyperpunitiveness" (Herbert & Brown 2007: 755) by blacklisting them from selective educational institutions, workplaces, and community spaces. The following sections, through instances of different garbo-curricula and garbo-pedagogies, discuss how neoliberal education systems operate.

Cherry-picking

In 2014, when I was teaching in a higher secondary school in Varanasi, one day, a class eight student named Ayush (name changed) approached me with an issue. He shared that most humanities subjects are boring because they are all about dictation and taking notes, unlike science and mathematics. No experiments or first-hand learning experiences take place. When I told him that he was wrong and there was enough scope to be creative and participative while learning subjects like history, literature, and other humanities subjects, he was not very convinced because the designated teachers in his class, instead of telling them stories and explaining the different contexts of the historical and literary narratives, forced the students to take down notes for examinations. While perusing the notes, I observed that based on the previous years' question papers, the notes had been cherry-picked from different parts of the books without engaging with the narratives holistically.

After Ayush's concern, I started conversing with other students from different classes, and many shared the same problem. Shalini (name changed) from class nine shared that history and English literature were her favourite subjects at one point. However, the dictation-note-making pedagogical approach of the teachers in the school had made her lose interest in these subjects. This myopic pedagogical attitude is prevalent in various schools and higher educational institutions across India. The teachers defend it as an ideal preparatory process for competitive examinations and job-oriented courses. According to many teachers, detailed reading and analysis are often a 'waste' of time and do not give the necessary knowledge that the students require to carve a successful career path. The celebration of this problematic pedagogical attitude leads to the "cultural asphyxiation" (Odora-Hoppers and Richards 2012: 8) of the learners and generates a forever-damaging "epistemological vacuum" (Odora-Hoppers and Richards 90). Individuals who do not conform to such attitudes are reduced to what Kevin Bales problematizes as "disposable people" (Bales 2004), who are regarded as having no intellectual worth.

As a part of the cherry-picked curricular and pedagogical process, the self-realizing capability of the learners about the intellectual and aesthetic values of different subjects and knowledge systems is uprooted and replaced by the dictatorial mandates of future-building that have been chalked out by the teachers, educational institutions, family members and societies on behalf of the learners. Consequently, the learners suffer a "confused consciousness and identity crisis" (Ndlovu-Gatsheni 2013: 178). In 2022, I had a chance to engage in a personal conversation with Ayush. He lamented that he is a successful engineer according to his parents' and school teachers' terms and conditions. Still, he is not satisfied with his work life because he envisioned a career as a fashion designer, which he was not allowed to pursue. This is how the culture of cherry-picked garbage pedagogies and curricula in India has made the spontaneous and free-flowing processes of intellectual development get "shrunk-wrapped ... into a corner in which the only echoes that resound" are a garbocratic heritage of impulsive and implosive knowledge-garbage laboratories (Odora-Hoppers and Richards 2012: 65).

Denying

The cherry-picked garbo-curricula and garbo-pedagogies are further institutionalized and systematized through the cultures and philosophies of denial. The practice of denial in educational institutions in India is nurtured by predetermining the ideal subject disciplines for the learners based on their percentage of marks. For instance, students who score very high marks (85 percent and above) in the intermediate are expected to pursue a career in engineering, science, and technology. The students who score a mediocre percentage of marks (between 75 percent and 85 percent) are expected to pursue a career in commerce. The students who score average and below average marks (65 percent or below) are expected to pursue a career in the humanities because, per the established perception, it does not require a critical mindset (Petriglieri 2018). Irrespective of the grades, if individuals wish to choose a career path different from the abovementioned established norms, they are structurally denied by the educational institutions through demoralizing and discouraging admission policies. In 2020, Ayesha (name changed), a class 12 student from a premier school in Kolkata, shared that despite her interest in the field of humanities, she was forced to choose the field of sciences by her parents and school teachers because she got high grades in the high school exam. On a similar note, Rajnish (name changed), a class 11 student from Kolkata, shared that he is denied essential financial and infrastructural assistance at home and school and is subjected to habitual mockery because, despite low grades, he chose to pursue a career in sciences, which is his subject area of interest.

This wasteful, arrogant, divisive, and derogatory practice of denial leads to the "objectification/thingification/commodification" (Gatsheni 2015: 490; also see Dey 2021, 2022a) of learners' psyche and intellect to such an extent that after a certain period, they regard their compromised pursuance as a usual pattern of knowledge development and inject the same forms of toxic knowledge making patterns within the consequent generations, like the tentacles of an octopus (Nkrumah 1965: 9). The practice of denial psychologically captures, suffocates, silences, and castrates the self-motivating and self-realizing capabilities of the learners. The impetuous technocratic designs of India's present-day higher education system can be cited as another relevant example. Though India is rapidly expanding its digital networks by initiating internet and telephone networks across rural and urban areas of the country, the privatization of these sectors in the name of providing better and quicker services has been escalating the digital divide in the country. Privatization has increased the accessibility costs of internet and telephone services, and as a result, many people are socioeconomically denied these services (PTI 2021; India Development Review 2023). During Covid-19, this divide was sharply visible. Many youngsters committed suicide because their parents could not afford a digital device and internet facility to avail of online classes in schools and colleges (Shrinivasa 2020; Koshy & Srinivasan 2020; Nair 2021). These experiences of denial have given birth to the 'anxiety of invalidation' amongst the learners, who, through the consistent experiences of rejections and impositions, are pushed to a "promised land" (Mignolo 2007: 450) of pseudo-happiness and pseudo-success, which gradually turns out to be suffocating, torturous, and nightmarish.

Ignoring

They deny gains further impetus through the curricular and pedagogical phenomenon of ignorance. Besides what has been discussed about the conscious performance of ignorance in the previous chapter, the phenomenon of ignorance resembles an assemblage of wastes – lively and lifeless, vibrant and silent, functional and non-functional, mobile and immobile, and manageable and disruptive at the same time. Ignorance of the student's educational and career interests occurs through similar patterns, where contradictory attitudes interplay, interweave, and coexist like piles of discards – huddling, cuddling, and intertwining. When the practices of cherry-picking and denial fail to provide the desired results to the neoliberal academic setup, the educational, research, and career interests of the learners are troubled, regulated, and manipulated by ignorant curricula and pedagogies that are driven by the values of vandalistic epistemologies, narcissistic ontologies, and ostracized psychologies.

Already, I have discussed in previous chapters and this chapter how selective episodes of Islamic, Dalit, northeastern, and Indigenous histories are eroded from bodies, psyches, intellects, and archives. In continuation, the methodology of curricular and pedagogical ignoring is executed through the construction of physical and ideological compartments within societies, families, and educational institutions, instructed by specific 'fit in' policies. These 'fit in' policies dictate that learners should not question teachers and elders because, by default, they are all-knowing and all-pervading entities;

they should not critique the subjects, ideas, and arguments taught in the textbooks because the education system has recommended the books for the welfare of the learners, so they should not be questioned; should not follow their own choices of academic and career goals and must unquestioningly abide by what the school and family institutions expect them to do, as schools and families can never be wrong; and should not be vocal about sociopolitical corruption and catastrophes, because causing collateral damages and indulging in unethical activities are the celebratory mantras of sociopolitical success.

To further establish the context of these arguments, let us reflect on how the ad-hoc faculty members were terminated from the Department of Sociology at Indraprastha College for Women, New Delhi, in September 2023. The department was founded in 2017 by a group of ad-hoc faculty members, mostly Muslims, Dalits, and residents of Northeast India, and within a brief period, it gained significant attention and appreciation for its critically diverse (Steyn 2015) curricular and pedagogical approaches. These approaches thoroughly interrogated the dumb, wasteful, and uncritical attitude of right-wing neoliberal knowledge systems in India. As expected, with time, the department posed a serious threat to the corrupted designs of the BJP government. As a result, in 2023, all the ad-hoc faculty members were removed and replaced by underqualified candidates, who during the selection interviews, displayed close associations with the BJP and convinced the panel of their sycophantic support for right-wing neoliberal ideologies. Furthermore, even before the interview results were officially declared, the selected candidates were being congratulated on Facebook for their selection, which exposed the corrupt ways in which the selection process was conducted. However, except for a few tweets and Facebook posts from the victims and their well-wishers, the incident went largely unreported. Except for *The Tribune* and *The Telegraph India*, no Indian media houses covered the incident. The concerns and grievances were systematically ignored because, culturally and intellectually, Dalits, Muslims, and northeast Indians are not seen as 'valuable' enough to be advocated for. Is this incident any less intense and nauseating than stinking heaps of garbage? Were the ad-hoc faculty members treated as anything more than discarded objects? Historically, their social, cultural, and existential spaces have been perceived as experimental grounds for politico-cultural narcissism, vandalism, and ostracization. They have always been taken for granted as a reference point for explaining incivility and primitiveness. They have always been footnoted and invisibilized within the discourses of mainstream cultures in India. It is no different today – scarred, discarded, distorted, ignored, and volatilized.

Silencing

What happens when, despite all the violence, scars, scratches, and abuse, these "volatile bodies" (Groz 1994) continue to resist and question? Like distorted trash cans, broken plastic toys, overflowing waste zones, and littered garbage bins, they are incinerated and erased through the pedagogical practices of silencing. As observed in the previous section, the curricula and pedagogies of silencing are curated first by identifying the bodies, psyches, and voices that have been persistent, resistant, and vocal against the mis-creations of neoliberal knowledge systems. Once identified, they are physiologically and ideologically penalized by erasing them, their documentation and archives, and replacing them with a triumphalist "classificatory system" (Gordon 2023a: 25) of ruthless, wasteful, reductionist, perverted, and contextless ideologies. After replacement, the corrupt modes of knowledge production are bubble-wrapped with pseudo-ethical and pseudo-moral values maintained through compromised principles (Gordon 2023: 30) of sanity, security, and discipline.

This is why, when, in 2020, right-wing masked hooligans rampaged the campus of Jawaharlal Nehru University and beat up students for consuming non-vegetarian food during Ram Navami,[2] it was 'silently' defended by the BJP ministry as an act of protecting the traditional values of India (Baruah and Sinha 2022); in 2019, when students of Jamia Millia Islamia University were ostracized and brutalized for protesting against the

---

2   Ram Navami is a Hindu festival that celebrates the birth of Rama, a revered Hindu deity who is considered the seventh avatar of Vishnu.

Citizenship Amendment Act, the BJP government 'silently' defended the brutality as an act of security (David 2021); and in 2022, when civil service aspirants were beaten up in New Delhi for demanding "an extra attempt at the UPSC exam because many could not appear for it due to COVID-19," the violence was defended as necessary to maintain discipline (Sukla and Dutta Roy 2022). These justifications create a "neurotic atmosphere of avowed disavowal" (Gordon 2023b: 35), where the ethicality of unethical actions prevails over justice and reason, as could also be seen in the cases of Sabyasachi Das and Gilles Verniers of Ashoka University, who were 'silently' forced to resign. Like the silence, malleability, viscosity, porosity, and invasiveness of sewer waste, which can flow in multiple directions uncontrollably, the toxic pedagogies and methodologies of silencing flow incessantly, invasively, and silently into educational institutions, streets, homes, dreams, blood, veins, consciousness and subconsciousness of individuals.

Silence is not entirely silent in its nature. It is lethal, silently vocal, and vocally silent. Like the Foucauldian panopticon and Deleuzian rhizome, silence stealthily watches, listens, surveils, records, and controls the knowledge-making patterns of humans through body language, cameras, censorship, and "misanthropic consciousness" (Gordon 2023c: 53). The central aim of misanthropic consciousness is to discard humanity in two directions – "the powerful are raised above the level of being human, and the least socially powerful is situated below it" (Gordon 53). The performatives of silencing create a tension between an "uncritical attitude towards favoured" views and "displeasing evidence" (Gordon 54), and this tension is consumed, internalized, and embedded through rhizomatic ignorance and "apophatic listening" (Samuelsson and Ness 2019).

Garbage, Garbage Everywhere, Not a Space to Think …

The aspects of garbo-curricula and garbo-pedagogies discussed in this chapter are not restricted to four-walled classrooms. They are spread across every habitual place and space of existence. The garbage-knowledge laboratories of schools and higher educational institutions teach not only through textbooks, chalk, dusters, and smart boards but also through iron rods, tear gas, and guns on university campuses and in the streets. When questioned, like discarded waste, the garbo-curricula and garbo-pedagogies intersect, polymerize, intertwine, interpellate, permeate, and multiply through the physical and ideological membranes of communities and societies in an apophatic and rhizomatic fashion. Even if we fail to see, identify, and listen to the garbo-curricula and garbo-pedagogies, they can apophatically listen to us and multiply against us like a rhizome – silently expanding within, around, and beneath us in multidirectional, nonlinear patterns. They are everywhere. They have occupied and entangled every nook and corner of our being and becoming.

Today, India has reached an existential and ecological standpoint where every ideological approach towards governance, politics, economy, education, and spirituality appears as garbage in form, formulation, and action. Knowledge can now be produced in contemporary India only through wasteful dehumanization, destruction, brittleness, and erasure attitudes. In time, the normalization of these attitudes mutates individuals into garbo-citizens who behave like a "junkyard Frankenstein" (Chen 2014). They are "surprising, bizarre, becoming, erratic and aberrant" (Ghosh 2021). They are stealthy. They are cunning. They are predatory. They are Kafkaesque. They can "present themselves, express themselves, transform themselves, let themselves be seen, produce themselves, spread themselves out, alter themselves, and conceal themselves" simultaneously (Ghosh 2021). They spell the onset of a tremendous human collapse, which will be multi-sensorial, transmedial (Ghosh 2021; Dey 2022b), planetary, and irretrievable.

Works Cited

Ayesha. (2020). "The Politics of Career Choices", Personal Conversation, Kolkata.
Ayush. (2014). "Dictation and Note-making in Schools", Personal Conversation, Varanasi.
Bales, K. (2004). *Disposable People: New Slavery in the Global Economy*. Berkeley: University of California Press.

Baruah, S. & Sinha, J. (2022). "JNU: Students Injured in a Scuffle Over 'Ram Navmi Pooja,' 'Non-Veg Food'", *The Indian Express*, <https://indianexpress.com/article/cities/delhi/jnu-violence-students-injured-7862961/>, 22 April 2023.

Bauman, Z. (2004). *Wasted Lives: Modernity and its Outcasts*. Cambridge: Polity Press.

Benowitz. J. M. (2009). "Reading, Writing and Radicalism: Right-Wing Women and Education in the Post-War Years", *History of Education Quarterly*, 49 (1), 89–111.

Boossabong, P. (2018). "Neoliberalizing Higher Education in the Global South: Lessons Learned from Policy Impacts on Educational Commercialization in Thailand". *Critical Policy Studies*, 12 (1), 110–115.

Chen, A. (2014). "Rocks Made of Plastic Found on Hawaiian Beach", *Science*, <https://www.science.org/content/article/rocks-made-plastic-found-hawaiian-beach>, Accessed 14 June 2018.

David, S. (2021). "'The World Moves On, but We're Stuck in the Same Evening': Remembering the Jamia Violence of 2019", *Newslaundry*, <https://www.newslaundry.com/2021/12/17/the-world-moves-on-but-were-stuck-in-the-same-evening-remembering-the-jamia-violence-of-2019>, Accessed on 21 December 2023.

Dey, S. (2021). "Pedagogy of the Stupid", *Philosophy and Global Affairs*, 1 (1), 22–45.

Dey, S. (2022a). "Pedagogy of Common Sense", *Tumultes*, 1, 255–272.

Dey, S. (2022b). "Pedagogy of Performative Silence", *Philosophy and Global Affairs*, 2 (1), 15–40.

Ghosh, R. (2021). "Desiring-Material: Plastic-Art and Affect-ability", *Minnesota Review*, 97, 53–76.

Ghosh, R. (2021). "The Plastic Controversy", *Critical Inquiry*, <https://critinq.wordpress.com/2021/02/04/the-plastic-controversy/>, 14 September 2023.

Giroux, H. A. (2023). "The Right Wing is Waging a Dirty War Against History and Education". *Truthout*, <https://truthout.org/articles/the-right-wing-is-waging-a-dirty-war-against-history-and-education/>, accessed on 15 January 2024.

Gordon, R. (2023a). "Reasoning in Black: Africana Philosophy Under the Weight of Misguided Reason". In R. Maart, S. Dey & L. Gordon (eds.), *Black Existentialism and Decolonizing Knowledge*, pp. 23–34. London & New York: Bloomsbury.

Gordon, R. (2023b). "Race in the Dialectics of Culture", In R. Maart, S. Dey & L. Gordon, *Black Existentialism and Decolonizing Knowledge*, pp. 35–50. London & New York: Bloomsbury.

Gordon, R. (2023c). "Racism as a Form of Bad Faith". In R. Maart, S. Dey & L. Gordon, *Black Existentialism and Decolonizing Knowledge*, pp. 51–55. London & New York: Bloomsbury.

Groz, E. (1994). *Volatile Bodies*. Bloomington: University of Indiana Press.

Herbert, S. & Brown, E. (2006). "Conceptions of Space and Time in the Punitive Neoliberal City", *A Radical Journal of Geography*, 38 (4), 755–777.

India Development Review. (2023). "India's Digital Divide: From Bad to Worse?", *Idr*, <https://idronline.org/article/inequality/indias-digital-divide-from-bad-to-worse/>, accessed on 15 February 2024.

Iovino, S. (2009). "Naples 2008, or, the Waste Land: Trash, Citizenship, and an Ethic of Narration", *Neohelicon* 36: 335–346.

Kalia, S. (2021). "Students Demand Northeast Culture, History Be a Part of School Textbooks to Counter Racism", *The Swaddle*, <https://theswaddle.com/students-demand-northeast-culture-history-be-a-part-of-school-textbooks-to-counter-racism/>, accessed on 23 August 2023.

Kikon, N. (2021). "NCERT Textbooks Need to Include a Chapter on Northeast to Educate 'Mainland' Indians." *The Print*, <https://theprint.in/campus-voice/ncert-textbooks-need-to-include-chapter-on-northeast-to-educate-mainland-indians/678621/>, accessed on 07 September 2023.

Koshy, S. M. & Srinivasan, C. (2020). "Unable to Join Online Classes, Kerala School Girl Commits Suicide: Cops." NDTV, <https://www.ndtv.com/india-news/coronavirus-kerala-girl-cant-attend-online-classes-amid-lockdown-commits-suicide-2239318>, accessed on 02 June 2020.

MHRD. (2020). *National Education Policy 2020*. New Delhi: Government of India.

Mignolo, W. D. (2007). "De-linking: The Rhetoric of Modernity, the Logic of Coloniality and the Grammar of De-coloniality", *Cultural Studies*, 21, (2–3), 449–514.

Mignolo, W. D. (2007). "Introduction: Coloniality of Power and De-Colonial Thinking", *Cultural Studies*, 21 (2–3), 155–167.

Mkandawire, T. (2011). "Running While Others Walk: Knowledge and the Challenge Africa's Development", *Africa Development* 36 (2), 1–36.

Mpofu, B. & Ndlovu-Gatsheni, S. J. (2020). "Introduction: The Dynamics of Changing Higher Education in the Global South". In B. Mpofu & S. J. Ndlovu-Gatsheni (eds.), *Dynamics of Changing Higher Education in the Global South*, pp. 1–13. New Castle Upon Tyne: Cambridge Scholars Publishing.

Nair, A. (2021). "Maharashtra: Unable to Attend Online Classes, Class X girl Hangs Self in Akola". Times of India, <https://timesofindia.indiatimes.com/city/pune/unable-to-attend-online-classes-class-x-girl-hangs-self-in-akola/articleshow/81890737.cms>, accessed on 04 April 2022.

Nawani, D. (2014). "North-east Indians and Others: Discrimination, Prejudice and Textbooks", *Economic and Political Weekly*, 49 (24), 19–21.

Ndlovu-Gatsheni, S. J. (2013). *Empire, Global Coloniality and African Subjectivity*. New York & Oxford: Berghahn Books.

Ndlovu-Gatsheni, S. J. (2015). "Decoloniality as the Future of Africa", *History Compass* 13 (10), 485–496.
Nkrumah, K. (1965). *Neocolonialism: The Last Stage of Imperialism.* New York: International Publishers.
Odora-Hoppers, C. & Richards, H. (2012). *Rethinking Thinking: Modernity's "Other" and the Transformation of the University.* Pretoria: University of South Africa Press.
Ogachi, I. O. (2011). "Neo-Liberalism and the Subversion of Academic Freedom from Within: Money, Corporate Cultures and 'Captured' Intellectuals in African Public Universities", *Journal of Higher Education in Africa*, 9 (1 & 2), 25–47.
Petriglieri, G. (2018). "Business Does Not Need the Humanities – But Humans Do", *Harvard Business Review*, <https://hbr.org/2018/11/business-does-not-need-the-humanities-but-humans-do>, accessed on 11 November 2023.
PTI. (2021). Experts Say, "Digital Divide Still a Challenge in Remote Teaching, Learning". *Times of India*, <https://timesofindia.indiatimes.com/world/rest-of-world/digital-divide-still-a-challenge-in-remote-teaching-learning-say-experts/articleshow/87700565.cms>, accessed on 23 October 2023.
PTI. (2023). "It's Insulting to Call 'Adivasis' as 'Vanvasis': Sharad Pawar Targets BJP". *Outlook India*, <https://www.outlookindia.com/national/it-s-insulting-to-call-adivasis-as-vanvasis-sharad-pawar-targets-bjp-news-269517>, accessed on 12 March 2023.
Rajnish. (2020). "The Politics of Career Choices", Personal Conversation, Kolkata.
Samuelsson, M. & Ness, I. (2019). "Apophatic Listening", *Democracy and Education*, 27 (1), 1–5.
Shalini. (2015). "Dictation and Note-making in Schools", Personal Conversation, Varanasi.
Shrinivasa, M. (2020). "Mysuru: Unable to Get a Smartphone for Online Classes, Girl Commits Suicide." Times of India, <https://timesofindia.indiatimes.com/city/mysuru/mysuru-unable-to-get-smartphone-for-online-classes-girl-commits-suicide/articleshow/77615140.cms>, accessed on 16 August 2022.
Shukla, S. & Dutta Roy, D. (2022). "Watch: Civil Service Aspirants Beaten Up, Dragged Off Delhi Protest", *NDTV*, <https://www.ndtv.com/india-news/watch-civil-service-aspirants-beaten-up-dragged-off-delhi-protest-3627091>, 12 December 2022.
Steyn, M. (2015). "Critical Diversity Literacy: Essentials for the Twenty-first Century". In S. Vertovec (ed.), *Routledge International Handbook of Diversity Studies*, 379–389. New York & London: Routledge.

SECTION IV
Regulating

CHAPTER 6

Garbo-Citizenship: The Great Human Collapse

Downfalls: Tangential Crises

The downfalls have already begun. These downfalls are not only human but also other-than-human in scale. They are physical, ideological, and metaphysical, woven with incidents and stories that are rarely exposed or reported. The trajectory of the downfall of human civilization into garbo-citizens has been traced not only in this monograph but also in two preceding ones, namely *Green Academia* (2022) and *Performing Memories and Weaving Archives* (2023), of which this book serves as a continuation. Before exploring further how I have subtly and suitably portrayed the aspects of garbo-citizenship in my previous monographs, let us investigate the historical factors that contributed to the evolution of garbo-citizenship. This investigation invloves three distinct events from different spatiotemporal contexts: the annihilation of the natural environment by the British colonizers during the seventeenth and eighteenth centuries and its ongoing impact, a personal teaching–learning experience with a particular section of class nine in a higher secondary school in Varanasi in 2015, and the declaration of the Citizenship Amendment Act in 2019. Although these three incidents may appear completely different, they are connected by a common phenomenological thread of garbo-citizenry and garbocracy – intrusive, disgusting, destructive, lucrative, and alluring.

Incident I

When the East India Company arrived in India to expand trade and commerce, "thousands of square miles of forests were destroyed for timber, and land mined for coal, for Britain's maritime expansion, shipping industry, and to build and maintain the Indian railways" (Bindra 2017: 5). However, the process of destruction and elimination was selective and profit-driven. The colonizers "imposed on the country a vision of nature that viewed beasts as vermin" (Bindra 6) and unleashed a "veritable war against errant species" (Rangarajan 2001, 17). The war was declared only against the species that did not fit within the company men's colonial, commercial, and capitalistic profitmaking designs. Those species of plants, animals, and birds that were commercially viable were captured and segregated from their natural surroundings, transported to Europe, selectively bred, medically and genetically modified, scientifically patented, and then publicly displayed and sold worldwide as their own discoveries. The natural environment was treated no less than garbage heaps – strategically sifted, siphoned, processed, and vanquished.

Many capsules, powders, liquids, leaves, twigs, roots, and syrups that are consumed across the world today as Indigenous medicines were once dehumanized, demonized, and rejected by the European colonizers as outdated and unscientific so that later, they could be stealthily hunted, gathered, appropriated, and museumized as European discoveries. To naturalize and systematize violations of the natural environment, the British recruited the socioeconomically privileged class of Indians, such as the kings and landowners, and heavily compensated them for their assistance. Today, the process continues through the garbocratic policies of the government, who recruit money launderers disguised as industrialists, land mafias disguised as policymakers, and sycophants disguised as security personnel to implement ecologically sustainable growth in India. This is why, during Jairam Ramesh's tenure as Minister of Environment and Forests (2009–2011), when several industrial projects within and around forest areas and river basins were shelved for the conservation of the natural environment, Prachi Bhuchar and Devesh Kumar, reporters from *India Today*, designated Jairam's environmental protection strategies as "green terror" and an "obstructionist approach" towards development (Bhuchar and Kumar 2013). This is why, even after being denied forest clearance twice by the Forest Advisory Committee (FAC) in 2013, the Dibang Multipurpose Project was "given forest clearance" in September 2014 (Bindra 2017: 15). In 2016, a new waterway act named the National Waterways Act, "which marks over 100 of India's rivers as National Waterways" (Bindra 25), was enacted to carry cargoes across aquatic highways. Due to these initiatives, several land and marine species were pushed into a threatening situation. Still, all the resistance and calls for dismantling the plans went unheard because they did not conform with the garbageous developmental vision of the government. If the voices are heard, the cobwebs of corruption and depletion that have been expanded across the forests, mountains, rivers, and oceans within and beyond the country's borders will collapse.

Incident II

In 2014, while pursuing my Ph.D. from Banaras Hindu University, I taught in a higher secondary school. From the first day, I observed how students were herded into different classroom sections following a grade-based segregationist approach. In each class, section 'H' (the last section) was designated for students who did not perform 'well academically' according to the school's educational parameters. They were not only concentrated in one section physically but ideologically separated from the rest of the students. On one side, students in the 'H' section were not allowed to freely mix and interact with others for fear they might 'spoil' them, and on the other side, the rest of the students were also discouraged from intermingling with those in 'H' section. The school unofficially nurtured and implemented this strategy through various hidden surveillance tactics, like recruiting students and staff to stealthily check if 'H' section students were interacting with students from other sections; if they were sharing study notes; and if they had access to the same private tutors as other students.

So, the monitoring process took place not only within the school but also outside. If the students were found 'guilty,' they were corporally punished, which was against the ethical principles of the school. But, again, unofficially, the teachers were given a unique upper hand to execute corporal punishments against the students of the 'H' sections. The consistent dehumanization and demonization of a particular group of students resulted in adverse consequences. A point in time came when the students of these sections became resistant to such maltreatment, and in return, they mistreated the teachers, whosoever came to teach in their classes. It became a nightmarish experience for the teachers to teach in 'H' sections. However, the teachers failed to understand an essential aspect: that the solution did not lie in disillusioning, dissolving, and bruising the bodies and psyches of the students, but in "the ability to know and understand" (Gordon 2010: 32) and to try to identify the problems through their lens, rather than imposing generic, universally formulated, and preconceived problem-solving blueprints. With time, through repeated suspensions and blacklisting, the students were hurled and disposed of from the school like unwanted waste. Physically and ideologically, they were pushed into such a devastating state that many went into an unrecoverable state of depression and psychological sterilization.

In that class, many students were very productive and successful in extracurricular activities like sports, music, dance, theatre, art, painting, and others. They also represented their school in national and international events and earned several laurels. However, they were not deemed successful or valid enough to be part of the school. In 2023, I conversed with a few students from the 2014 batch who were blacklisted, and many could not manage to gain access to a decent education and career. This is more than just the case of one school. Several other educational institutions across the country follow a similar garbocratic route of knowledge production, where the paradigms of diversity, creativity, inventiveness, and holistic learning are stuffed and pigeonholed within exam papers and grade sheets, just like garbage stuffed inside disposal bins and cast out.

Incident III

In 2019, the government of India launched the Citizenship Amendment Act (CAA) as an amendment to the Citizenship Act of 1955 to allow persecuted minorities like the Sikhs, Christians, Hindus, Buddhists, Jains, and Parsees who entered India from the subcontinent on or before 31 December 2014 to be granted Indian citizenship. However, in reality, the CAA gave more preference to Hindus over every other community, and over time, it was realized that the CAA, like many others, is an eye-washing strategy to criminalize selective communities like Muslims and Bangladeshi immigrants. The CAA appears to be a contemporary deformed version of the flawed postcolonial Indian Dream that ensured "protection for equality rights, particularly pertinent in a community rife with caste and class-based discrimination;" guaranteed "secularism in a nation fraught with communal antagonism and violence;" and placed "responsibility on the State to promote education in a country having a literacy rate of approximately 18 percent at the time of independence" (Lalli 2020: 101), but only catered to a particular group of socioeconomically privileged communities, who happily volunteered to pursue colonial cargo cultism and hangovers, even after the judicial independence of India from the British.

The CAA serves as a derogatory extension of the Indian Dream by exterminating communities who have legal residential permits and accommodating illegal communities by issuing fake identity documents; targeting specific population groups whose ancestors arrived in India from Bangladesh and Pakistan; targeting vulnerable women and children and subjecting them to trafficking and sexual violence; demolishing the temporary residencies of the refugees; and in many other ways (Bhatt 2019). To take legal action against the falsely hunted convicts, the BJP has also constructed Nazi-styled (both architecturally and ideologically) detention camps in Assam, where reasonably classified migrants (Lalli 102), who do not qualify for the communally designed sociopolitical desires of the political party are being incarcerated and quarantined, far from the outreach of their families and societies. As a result, the CAA is seen as "acutely detrimental to the religious freedom and secular values of India" (Khan 2020: 2).

It is another monumental initiative by the BJP to systematize Hindumania in the name of restoring a fraudulent version of traditional and ancient pre-colonial India. Post-CAA India has experienced a phenomenal rise in anti-Muslim and anti-refugee violence triggered by political leaders, BJP goons, police, and residents (HRW 2020).

These three scattered incidents from different spatial, temporal, and contextual moments have one aspect in common: they have all contributed to the evolution of garbage citizenship in the country, which, like unwanted discards, are vulnerable, directionless, closeted, and indecipherable. The appropriation of the natural environment, corrupted governing policies, unethical blacklisting of students, the incarceration of refugees, and many other factors have collectively contributed to the growth of garbo-citizens and garbo-citizenship in India. The phenomenon of garbo-citizenship encompasses both the victims and the perpetrators. It does not spare anyone by fissuring through the physiological cracks and crevices of daily life. Though there could be various other contexts to trace the birth of garbo-citizenship in contemporary India, I have chosen these three particular incidents because, personally and collectively, they have played an essential role in shaping this project and the arguments on garbocracy in this book. In the following sections, I have oulined the perspectives in which I have addressed the aspect of garbo-citizenship in my previous monographs.

Garbo-Citizenship in Natural Green Spaces: *Green Academia*

The monograph *Green Academia: Towards Eco-Friendly Education Systems* (2022) outlines how the European colonial tradition of transforming "nature from a 'source' of knowledge into a 'resource' of hunting, gathering, extracting, and profitmaking" (Dey 2022b: 3) continues to be performed in contemporary times through "capitalist industrialization" (Watts 2004: 196) and "systematic extraction of desired economic products" (Sivaramakrishnan 1995: 7). The book exclusively focuses on the colonial/capitalist education systems in selective countries like India, Bhutan, New Zealand and Kenya and unpacks how the roots of the "modern/colonial capitalist world system" (Grosfoguel 2002: 71) lie in the mainstream curricular and pedagogical patterns that promote a 'one-size-fits-all' degree-based and certificate-based knowledge-building structure. Such a knowledge-building structure eventually leads to the evolution and growth of garbo-citizens and garbo-citizenship. The arguments have explicitly been justified through unfolding the problematic methodologies of assessing student and faculty performances in educational institutions, such as a "mania for assessments," "flawed methodologies of student and faculty evaluation," "conversion of students into customers and consumers," "hierarchy of the English language," and "carryover of colonial texts" (Dey 11–12).

In the schools and higher educational institutions in India, the "system of business principles and statistical accountancy" (Mbembe 2016: 31) compels individuals to focus more on administrative work and remain distracted from the "contextually productive academic and research engagements, which can ecologically contribute to the sustainable development of a society" (Dey 11). As a result, to be eligible for promotions and financial incentives, researchers, students, and faculty have to produce quick, faulty, and plagiarized publications, which are systemically acknowledged by evaluating faculty and students on "number of publications, several conference papers presented, numbers of committees served on, number of courses taught, numbers of students processed in those courses, numbers of advisees, quantitative measures in teaching excellence, etc." (Mbembe 31).

The marketized and industrialized education systems have pushed individuals into epistemological and ontological states of "meaninglessness, powerlessness, and belongingness" (Clark 1959: 849), which ultimately leads to the transition of humans from 'human citizens' to 'garbage-like citizens' or 'garbo-citizens.' The production of garbo-citizens is maintained through a "death project" (Suarez-Krabbe 2012: 338) of "anti-nature and ego-centric methodologies of capitalistic education," which are curated through "textbooks written by a white and white-privileged author through the whiteboards and projectors in the classrooms; through grading and certificate distribution businesses; and the laboratories ..." (Dey 23). This is why, as elaborated in the previous section, the destruction of the natural environment in the name of development, the demonization of students in the name of academic growth, and the criminalization of citizens in the name of law and order can be deployed in such normative and celebrative ways, without concerns, realizations, and regrets.

Another major factor in generating garbo-citizenship in contemporary times is the lack of understanding of India's interwoven, transborder, and transcultural existence of societies, cultures, and economies. The monograph *Performing Memories and Weaving Archives: Creolized Cultures Across the Indian Ocean* (2023) discusses this issue.

Transoceanic Creoles and Garbo-Citizenship: *Performing Memories and Weaving Archives*

One of the many aspects that the book engages with is the "problem of retrieval" (Gudavarthy 2019: 191) of selective historical narratives that fit into a predetermined, communally enclaved, and Hindu-centric delineation of precolonial India. In the context of the South African Indians in South Africa and the African Indians in India, the book discusses how diaspora communities' socially, culturally, and historically creolized existence is resisted as impure, ill-cultural, and alienating. Primarily, in the context of the Afro-Indian diaspora in India, except for a few racially and geopolitically half-read and stereotyped narratives published in journals, newspapers, books, and magazines, their histories, cultures, and existence are hardly known across India and the world.

The process of not knowing is deliberate, systematic, institutional, and racially connotative. No textbooks in schools or higher educational institutions engage with the transoceanic, creolized, multirooted, and multicultural histories of the Afro-Indians because, according to the mainstream knowledge-making attitudes, they are not relevant enough (Dey 2023: 7; also see Dey, Maity & Srivastava 2024). The existence of the community is only validated (or pseudo-validated) when it becomes necessary for the Indian government to depict a false picture of sociocultural diversity in front of international delegates by forcing them to dress, dance and sing in ways that are colonial/globally/historically perceived as markers of authentic Africanness (Dey 2021) – an "ontology of reductive sameness" (Ngwena 2018: 1) through which Africans and African cultures and histories are interpreted as sterilized homogeneous components. As a result, because of their hair patterns, facial structures, skin colours, and other anatomical characteristics, they are habitually harassed in schools, public roads, and workplaces. Like waste, they have been squashed, trashed, decimated, and reduced to garbo-citizens.

Another reason why the histories and cultures of the Afro-Indians in India remain deliberately unrealized is that they are Muslims, and the garbage Islamophobic intentions of the BJP government are not unknown. The reduction of the Afro-Indians to garbo-citizens has reduced pluralized Indian histories, religions, regimes, and politics to "those captured in the mainstream of Brahminic Hinduism, which is a partial and misleading presentation of India" (Rathore 2017: 77). Concerning the multifarious social, economic, cultural and political transitions, India is currently experiencing widespread "contextual communalism" (Gudavarthy 238). Contextual communalism "emerges from the complexities of its diversity and political construction of communities in terms of being the majority and minority groups, wedded to the electoral process and significance of political power linked to the discourses of nationalism" (Gudavarthy 238).

The diminution and transmogrification of Afro-Indianness into garbo-citizenship have also erased significant episodes of India's historical and cultural evolution that document real-life experiences and folklores of transoceanic, transracial, transcultural, interreligious, and human-more-than-human interactions. These dynamics of producing garbo-citizens and garbo-citizenship continue to be interrogated and critiqued in this book and have been further investigated through wider analytical lenses of garbo-imperialism, garbo-knowledge, garbo-power, and garbo-beings. This book is a paroxysm of the loss of human, humanness, humanity, and humility, along with critiquing, arguing, problematizing, and expressing frustration and anger against an impending garbocracy.

The Machiavellian Garbage: Towards an Ever-expanding Barrenness

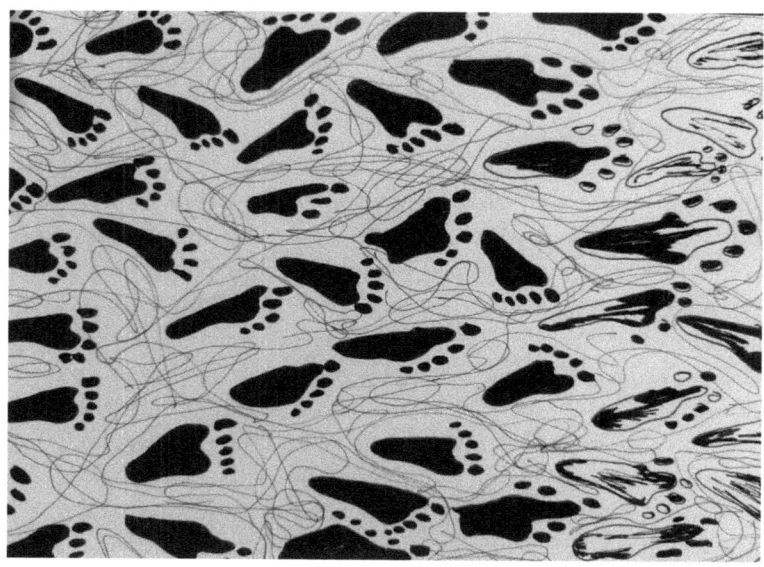

'The Machiavellian Garbage'. Artwork by Sayan Dey.

Garbo-citizens and garbo-citizenship in India have entrusted the phenomenon of garbage with Machiavellian characteristics, pushing societies, communities, and cultures towards a state of ever-expanding existential barrenness. The Machiavellian garbage operates in the following forms:

a. *Neurotic Garbage*: The garbageous ideologies through which societies are governed, cultures dictated, and knowledge produced in entanglement with physical waste, cause various forms of nervous breakdowns and mental illnesses. As discussed in the chapters, garbage, garbaging, and garboglomerates are devious, deceptive, invasive, and hauntological. Besides physically impacting humans, they also generate neurotic impacts. These neurotic impacts are visible through the systematization and normalization of violent, exploitative, and corruptive governance, educational, commercial, and sociocultural policies across generations, which go unquestioned. In a neurotic garbage state, questioning these problems is deemed abnormal and unethical. There is a consistent failure to understand the "life of the mind" (Marsden 2023: 114).

Neurotic garbage also creates situations of intellectual sterilization, compromising the cultures of free thinking, critical understanding, and outspokenness with values of monopolism, narcissism, and dictatorship. Consequently, anti-corruption campaigns are suppressed by the state through guns and rods; journalists reporting on government corruption are imprisoned with fabricated charges or publicly hunted down in broad daylight; educational institutions that promote genuine critical and intellectual approaches to knowledge dissemination are shut down; research projects that do not serve the communal, political, racial, and cultural propaganda of the current government are defunded; and academic disciplines and degrees that do not generate sufficient revenues and profits are defunctionalized.

b. *Schematic Garbage*: Neurotic garbage doesn't just exist on its own, but like particles and molecules in discarded waste, it polymerizes into other existential dimensions, such as schematic garbage. Schematic garbage is diagrammatic and architectural in form. It can be seen in ideologically self-centric, emotionally exclusionary, and socioeconomically hierarchical, cobwebbed residential complexes in Indian urban areas. These urban residencies exist not only as physical entities but also strategically and schematically give birth to high-walled, high-rise, exclusionary "cartographies of power" (Clark 2003: 8), within which selective bodies, psyches, and ideologies are socioeconomically invalidated and prohibited from entry because they are not socially, culturally, or economically viable. For example, in gated complexes, individuals such as street food sellers, vegetable vendors, and hawkers are not allowed entry because, according to preconceived social dynamics, they are from lower cultural backgrounds and, therefore may not be deemed respectable. The items they sell may also be perceived as of inferior quality.

However, house cleaners and caretakers from the same cultural backgrounds as the vendors and hawkers are permitted entry because their bodies and psyches are readily available for physiological and psychosocial exploitation in the name of economic empowerment and cultural upliftment. In return for fulfilling the demands and desires of the residents, they are compensated negligibly. This is how schematic garbage hierarchizes and schematizes urban residential complexes as "spaces of modern civility" (Bayly 2023: 40) and "moralizing geohistories" (Bayly 43), where entry and exit hinge on class status and money-making agendas rather than on values of humanity, mutual respect, and holistic accessibility. Due to the regulatory impact of schematic garbage, existential issues such as loneliness, harassment, abuse, financial debts, and other challenges often go unreported because they are either considered sociocultural stigmas or not severe enough to warrant attention. The heaps of discarded garbo-citizens that neurotic and schematic garbage generate are further multiplied by haptic and prognostic garbage.

c. *Haptic Garbage*: Another Machiavellian form in which garbage impacts human civilization is haptic garbage, which includes mobile phones, smart televisions, smart boards, smart cities, intelligent environments, laptops, notebooks, and other so-called comfort gadgets. These objects present themselves as valuable and life-changing, but in reality, they are destructive. In the name of development, globalization, inclusivity, diversity, and modernity, these objects increasingly narrow, restrict, disembowel, and fracture the collective, cooperative, and co-existential values of humans and other-than-humans. Haptic garbage motivates humans to touch, feel, cuddle, internalize, embed, shovel, and consume garbage objects in a hypnotically celebrative manner, functioning like a contagion and infecting others. This explains why people have more faith in rumours rather than well-researched information; rely more on communicating through mobile applications than face-to-face interactions; favour semiotic expressions like GIFs and emojis over real-life spontaneous expressions; and depend on AI applications for facts, figures, ideas and statistics, thereby reducing the human brain to a non-functional junkyard.

When the above-mentioned haptic garbage is installed over cupboards, walls, in bedrooms, drawing rooms, workplaces, and classrooms, it offers ample promises and opportunities in software, hardware, storage spaces, easy lifestyles, visual clarity, and aesthetic qualities. Yet, the moment they are operated, users become entrapped within a set of fixed and inescapable emotions, colour schemes, visuals, narratives, and ideologies, that falsely pose as genuine, realistic, and beneficial to well-being. Like the obstructive, obtrusive, invasive, and seductive nature of openly discarded objects, haptic garbage is seductively appropriative, imposingly manipulative, restrictively expansive, exclusively inclusive, and fictionally realistic.

d. *Prognostic Garbage*: Machiavellian garbage can also act as a fortuneteller in the form of prognostic garbage. Every form of physical, ideological, and psychological garbage discussed in this book, in their movements, multiplications, displacements, destructions, and concoctions, foretell a highly threatening, insecure, impulsive, and garbageous future filled with irreparable physiological and ideological barrenness across bulldozed political systems, educational structures, sociocultural practices, emotional values, and natural environments. This barrenness intends to be infectious, interminable, unpredictable, and implosive. Like a garbocracy, it seeks to be simultaneously visible and invisible, tangible and intangible, and labyrinthine and unravelled. The barrenness is imperialistic and compartmentalizing, dividing bodies, minds, emotions, and psyches into vulnerable fragments susceptible to disruption and erasure. Barrenness is also genetic, transfusional, and trans(in)fusional (Ghosh 2021b) in forms and functions, as it motivates, manipulates, and compels generations of humans to adhere to the cataclysmic modes of being and becoming without questioning or critical intervention.

The "chemical-organic-material-experimental-global-structural outgrowth" (Ghosh 2021a: 65) of garbocracy into polymerized and multiplicative garbocratic futures is only a matter of time. A garbocratic future provokes us to ask many questions. What will a garbocratic future look like? How will it function visibly, invisibly, physically, and ideologically? What will be the parameters of garbocratic governance? Are there ways to resist garbocracy? The answers to these questions are camouflaged in the present through vibrations, silence, and despair.

Works Cited

Bayly, S. (2023.) "Mapping Time, Living Space: The Moral Cartography of Renovation in Late-Socialist Vietnam." In J. Copeman et al. (ed.), *Anthropology of Intellectual Exchange: Interactions, Transactions, and Ethics in Asia and Beyond*, pp. 39–66. New York & Oxford: Berghahn Books.

Bhatt, S. (2019). "Why India's New Citizenship Plans Are Stirring Protests", *Reuters*, <https://www.reuters.com/article/india-citizenship-explainer-idINKB N1YR0W7>, accessed on 23 December 2023.

Bhuchar, P. & Kumar, D. (2013). "Green Terror: Outdated Environmental Laws and Inflexible Ministers Strangle Indian Economy", *India Today*, <https://www.indiatoday.in/india/story/green-terror-outdated-environmental-laws-and-inflexible-ministers-strangle-indian-economy-117900-2012-10-04>, accessed on 19 December 2023.

Clark, J. P. (1959). "Measuring Alienation Within a Social System", *American Sociological System*, 24 (6), 849–852.

Clark, K. (2003). "Socialist Realism and the Sacrifice of Space". In E. Dobrenko & E. Naiman (eds.), *The Landscape of Stalinism: The Art and Ideology of Society Space*, pp. 3–18. Seattle: University of Washington Press.

Dey, S. (2021). "'Killing with Kindness': Political Icons, Socio-cultural Victims: Visual Coloniality of the Siddis of Karnataka, India". In B. Karam & B. Mutsvairo (eds.), *Decolonising Political Communication in Africa: Reframing Ontologies*, pp. 91–108. New York & London: Routledge.

Dey, S. (2022a). "Eco-Friendly Academic Systems: A Journey to the Roots." *Green Academia: Towards Eco-Friendly Education Systems*, pp. 23–66. New Delhi: Routledge India.

Dey, S. (2022b). "Introduction: Why Green Academia?" *Green Academia: Towards Eco-Friendly Education Systems*, pp. 1–22. New Delhi: Routledge India.

Dey, S. (2023). *Performing Memories and Weaving Archives: Creolized Cultures Across the Indian Ocean*. New York: Anthem Press.

Dey, S., Maity, T. & Srivastava, T. (2024). "Gender Empowerment in Transoceanic Feminine Folklores and Shrines: A Kin Study of Siddi Women's Participation in Mai Misra Worship in Gujarat, India", *Journal of International Women's Studies*, 26 (1), 1–15.

Ghosh, R. (2021a). "The Plastic Turn", *Diacritics*, 49 (1), 64–85.

Ghosh, R. (2021b). *Trans(in)fusion: Reflections for Critical Thinking*. New York & London: Routledge.

Gordon, L. R. (2010). "A Pedagogical Imperative of Pedagogical Imperatives", *Thresholds in Education*, XXXVI, 27–35.

Grosfoguel, R. (2002). "Colonial Difference, Geopolitics of Knowledge, Global Coloniality in the Modern/Colonial Capitalist World-System", *Review (Fernand Braudel Center)*, 25 (3), 203–224.

Gudavarthy, A. (2019). *India After Modi: Populism and the Right*. New Delhi: Bloomsbury India.

HRW. (2020). "'Shoot the Traitors': Discrimination Against Muslims Under India's New Citizenship Policy", *Human Rights Watch*, <https://www.hrw.org/report/2020/04/10/shoot-traitors/discrimination-against-muslims-under-indias-new-citizenship-policy>, accessed on 20 July 2023.

Khan, T. (2020). "The Citizenship Amendment Act, 2019: A Religion Based Pathway to Indian Citizenship", *SSRN*, <https://papers.ssrn.com/sol3/papers.cfm?abstract_id=3665743>, accessed on 27 August 2022.

Lalli, J. S. (2020). "Communalisation of Citizenship Law: Viewing the Citizenship (Amendment) Act 2019 through the Prism of the Indian Constitution", *University of Oxford Human Rights Hub Journal*, 3 (1), 95–122.

Marsden, M. (2023). "Intellectual Exchanges in Muslim Asia: Intersections of History and Geography". In J. Copeman et al. (ed.), *Anthropology of Intellectual Exchange: Interactions, Transactions and Ethics in Asia and Beyond*, pp. 113–137. New York & Oxford: Berghahn Books.

Mbembe, A. (2016). "Decolonizing the University: New Directions." *Arts & Humanities in Higher Education*, 15 (1), 29–45.

Ngwena, C. (2018). *What is Africanness? Contesting Nativism in Race, Culture and Sexualities*. Pretoria: Pretoria University Law Press.

Rangarajan, M. (2001). *India's Wildlife History: An Introduction*. Ranikhet: Permanent Black.

Rathore, A. S. (2017). *Hegel's India: A Reinterpretation, with Texts*. New Delhi: Oxford University Press.

Sivaramakrishnan, K. (1995). "Colonialism and Forestry in India: Imagining the Past in Present Politics", *Comparative Studies in Society and History*, 37 (1), 3–40.

Suarez-Krabbe, J. (2012). "Identity and Preservation of Being", *Social Identities*, 18 (3), 335–353.

Watts, M. J. (2004). "Antinomies of Community: Some Thoughts on Geography, Resources and Empire", *Transactions of the Institute of British Geographers*, 29 (2), 195–216.

CHAPTER 7

Digital Discardscapes, Digital Infrastructures, and Garbocratic Futures

"Can we live together?"

(Latour 2005: 254)

This question posed by Bruno Latour is not just existential but phenomenological as well. The possibility of finding ways to live together also prompts us to ask: what does living together entail? Who can live together, and who cannot? In the context of digital infrastructures, digital discards, and garbocracy, these questions serve as the foundation of this chapter. The phenomenological dimensions of living together within specific digital infrastructures and through particular patterns of digital discards have been elaborated in this chapter through various ethnographic research. The concept of garbocracy reveals how wastes are not lifeless entities but function as a "planetary mark" (Armiero 2021: 2) of a new epoch where humans and wastes inseparably co-exist in entanglements and cobwebs, either out of choice or compulsion. Through conversations with waste collectors, the chapter discusses how digital infrastructures in India regulate the patterns of disposing of electronic waste, reflecting the varied class, cultural, and social statuses and positionalities of communities. The chapter also outlines the visual, economic, and cultural spectacularities that digital discards and infrastructures generate and how these spectacularities systematize the social, cultural, and economic divisions and hierarchies.

Smartness, Digital, and the Discards

A few years ago, in a so-called posh locality in Kolkata, I was walking past a garbage heap consisting mainly of half-burnt and broken hard drives, dismantled desktops, and unusable laptops. Although the heap did not appear as disgusting as foul-smelling waste, the sight of digital discards made me feel disgusted and suffocated. These feelings were further intensified by the fact that the disposers (people from surrounding homes and shops) and collectors, rather than discarding the electronic waste in an ecologically responsible manner, disposed of it in public spaces and burned it. The casual burning of discarded electronic objects is often considered innovative and spectacular. It is underpinned by the illogical, illusionary, and restrictive belief that only liquid and semi-liquid foul-smelling objects must be identified as discards, while the rest need not be. Also, another illusionary perspective generated by digital smartness is that possessing large quantities of electronic objects enhances the smartness and class status of individuals, which eventually causes "catch-all obsessive-compulsive disorder" or "disposophobia" (Humes 2012: 9). Besides accumulating massive volumes of electronic objects, individuals are often trapped in a pseudo-belief that unplanned and open disposal of electronic waste generates a spectacle, suggesting how specific communities and localities are digitally privileged and, in order to show off their privileges, do not hesitate to carelessly discard their excess digital gadgets out in the open.

Digital Discardscapes and Garbocratic Futures

'A Tower of Trash'. Artwork by Sayan Dey.

The condition of disposophobia with digital items and the unplanned disposal of electronic waste have become "one of the most accurate measures of prosperity in the twenty-first century" India (Humes 2012: 11), and the social and cultural intelligence of people in society is not measured by their capacity to think and act independently but by the extent to which they can function algorithmically like machines. For instance, children who can type, play online games, and decode passwords on phones and other electronic devices are considered more intelligent than those who interact physically with others, play outdoor games, and spend time with family elders. In many families, societies, and educational and work institutions, fast-paced, half-written, generic, and superficial information available on digital platforms and accessed through digital gadgets is seen as more valid than knowledge shared through in-person interactions. Today, these problematic mechanized approaches to building human consciousness and knowledge systems have led to the evolution of 'smart discards', particularly in urban areas. Smart discards can be interpreted as discarded electronic objects that serve to structure, systematize, dissipate, and celebrate social, cultural, and economic hierarchies in a spectacular way, where the array of discarded electronic objects signifies the socioeconomic statuses of residents in particular localities. The spectacle of disposing of electronic waste often intersects with the treatment of specific communities' discarded objects. Before proceeding further, it is essential to clarify that the purpose of critiquing the digital fetishism of individuals is not to undermine the importance of electronic gadgets and digital sources of knowledge in daily life but to highlight the dangers of hyper-dependence on digital gadgets and platforms, and their use as instruments for proliferating caste, class, and economic hierarchies.

The volume and types of electronic objects possessed and disposed of daily by individuals in urban localities[1] in India serve as socioeconomic signifiers exposing the social dynamics and power structures in society. In localities with high rates of electronic disposal, residents are typically socioeconomically comfortable and associated with hardware and software professions (Pendharkar 2018; Arya and Kumar 2020; Jadhav 2022). However, it is also crucial to note that users do not always dispose of electronic objects within their locality; rather these items are "transported, extracted, burned, pumped, emitted, and flushed" (Humes 2012: 14) to areas occupied by slums and squatter settlements, whose inhabitants are regarded as "literally full of shit" (Mathewson cited in Dlamini 2009: 132). This faecal imaginary portrays specific communities as "wasted humans" (Bauman 2004: 5), either due to their lack of access to diverse electronic objects or lack of interest in hyper-digital fetishism, making them individuals who are "not wished to be recognized or allowed to stay" (Bauman 2004: 5). Such a problematic attitude is an "inevitable outcome" of the spectacles of digital modernization and "an inseparable accompaniment

---

1   In order to adhere to the thematic direction of the workshop and eventual symposium, I have exclusively focused on urban areas in this article.

of [digital] modernity" (Bauman 2004: 5) in urban areas of India. The volumes and patterns of electronic disposal and the simultaneous classification of selective human communities as discards also shape and classify the spectacles of urban digital infrastructures, in forms such as high-speed network mechanisms and eye-catching electronic devices.

In this chapter, discussions on the impact of digital discards on digital infrastructures are centred on different localities in Kolkata and interwoven with the phenomena of social, cultural, and class hierarchies. This interwoven approach of analysis situates the theoretical and thematic arguments of the chapter within a broader historical framework of the capitalist modes of digital infrastructural developments, where the "consolidated amalgamation of elements, such as valuable, precious, hazardous and inert materials, altogether form a complicated structure and design" (Arya and Kumar 2020; see also Pinto 2008; Jayapradha 2015).

Garbo-infrastructures and Digital Discardscapes

> "… garbage has to be the poem of our time because garbage is spiritual, believable enough."
>
> (Ammons 2003: 62)

The observations by Archibald Randolph Ammons in the lines mentioned above directly speak to my experience of returning to my hometown in Kolkata from Bhutan in 2021. In 2021, after overcoming the exhausting technical, bureaucratic, and ethical challenges of COVID-19 lockdowns, when I arrived at my home in Kolkata from my workplace in Bhutan, I observed a heap of dry garbage just 50 metres away. Every time I passed by it, I could not help but look at it and find that the heap also consisted of various digital discards besides dry waste like discarded linens, pillows, and mattresses. I was also astonished that the professional cleaners who worked for the local municipality discarded most of the waste. When the people requested that they refrain from dumping waste in public places, they defended themselves by saying that the dump did not smell and, therefore, would not cause any harm to the locality. Later on, the residents in the locality launched a collective complaint against the unhygienic disposal of garbage, but the municipality did not pay any heed to it. After repeated complaints, the municipality chief drove away the residents by saying that he did not have time to look into petty matters and behaved as if co-living with discards is a standard and normative pattern of existence and refusing to do so is a crime. It was a petty matter for him because, at that time, the municipality was busy bulldozing a squatter settlement and choking a pond near my locality to build an internet network tower and improve digital connectivity, without which a heavy financial sponsorship from a private corporation would be lost.

This "evaluative (values/likes or dislikes)" (Teo and Loosemore 2010: 742) attitude of compromising with an ecological crisis and giving preference to a capitalistic luxury[2] generates a situation of "existential immobility" or "stuckedness" (Hage 2009: 97) which is a "very complex combination of cultural, social, institutional, political, organizational, ecological, historical and relational dynamics" (Luton 1996: 4). The stuckedness allows specific socially, culturally, and economically privileged communities to be hyper-mobile with excessive access to electronic objects at the cost of other communities. This, in turn, leads to the evolution of digital "discardscapes" (Lepawsky 2018: 15) that are "synthetic, unpredictable, and above all heterogeneous" in nature (MacBride 2012: 174). The discardscapes are "patchy, uneven, and not necessarily coherent" (Lepawsky 2018: 15) and are a "matter of concern" (Latour 2004: 237) for socially, economically, and hygienically vulnerable communities like waste collectors, who have to encounter the toxic digital discards for cleaning, burning, removing, and replacing daily.

As Rajib (name changed), a 25-year-old Dalit waste collector from the Belgachia region of Kolkata, observes: "We do not have any place in a

---

2   With respect to improving the internet network, the term 'capitalistic luxury' has been used because prior to engaging with the element of digital smartness in my locality, there are many basic and serious concerns regarding health and hygiene that need to be addressed, such as the lack of systematic disposal and processing of domestic wastes. In such a scenario, focusing on improving digital infrastructure without addressing these urgent issues becomes an act of 'capitalistic luxury.'

Digital Discardscapes and Garbocratic Futures 99

fast-paced digital world. We risk our lives to clean and burn digital waste, and instead, we hardly have access to digital facilities" (2022). He also adds: "Every time before elections, the government promises to improve the internet connectivity in our locality and provide us with free mobile phones so we can have quick access to health, medicine, and other necessary facilities. However, in the end, all the promises prove to be fake. Probably we are not 'smart' enough to have access to these spaces" (2022). The experience of Rajib shows how digital infrastructures are simultaneously inclusionary and exclusionary. On one side, the infrastructures need people like Rajib to "clean up" (Gray-Cosgrove 2015) the improperly disposed digital discards, exposing them to diverse health risks so that elite and digitally well-connected urban localities appear bright and clean. On the other side, they do not have access to many digital infrastructures because, as per the social, cultural, and economic systems in India, they are not meant to have any access and are expected to exist outside the realm of digital smartness (Mayaram and Agarwal 2022; Hullemane 2023). The discrimination reduces specific communities to the waste category, as lamented by Rupa (name changed), a Dalit female 28-year-old waste collector from the Shyambazar region of Kolkata. She says: "Because we are professional cleaners and spend our days removing trash from different localities, we are not treated better than garbage. So, why do garbage collectors need internet connections and mobile phones?" (2022) This provocative question shows that digital infrastructures in the urban regions of Kolkata in particular and India in general, under the exploitative and self-profiting capitalistic projects of digitization, celebrate the "exploitative, garbological and fast capitalist" systems (Ghosh 2022: 121) of development and modernity. In order to systematize the unequal production and access to digital infrastructures in urban India, the digital discards function "as an aspect of power" (Feenan 2007: 514), a closeted, claustrophobic, and spectacular space, and a borderline dictated by class and cultural hierarchies, which socioeconomically depleted communities must not cross.

As a result, one day, when Shyamlal (named changed), a 35-year-old Dalit waste collector from the Paikpara region of Kolkata, while collecting garbage from one of the houses, asked the owner if he could be provided with a second-hand mobile phone, the owner replied that as a cleaner he should concentrate on feeding his family rather than on digital luxuries (Personal Conversation 2022). The volume of digital discards denotes the extent to which communities and localities in the urban regions are digitally equipped and how a particular class of people, like waste collectors, have little to no access to digital infrastructures except for collecting and disposing of digital waste. This "politics of trash" (Strach and Sullivan 2023) that functions in collaboration with class, economic, and political hierarchies to regulate access and non-access to digital infrastructures, gives birth to garbo-infrastructures or garbageous infrastructures, which are toxic, disgusting, suffocative, and divisional like garbage heaps. Like us-and-throw objects, the profitmaking capitalistic intentions of digital infrastructures treat human societies as a "replacement market" (Pinto 2008: 65), where usable and discarded electronic objects continuously replace individuals because they are found incompetent to meet the rapid socioeconomic demands of contemporary times.

However, digitally marginalized communities like waste collectors try to resist their habitual segregation and disposition by exploring avenues of building tangential counter-digital infrastructures from the digital discards for non-discriminatory access and impact.

However, digitally marginalized communities like waste collectors try to resist their habitual segregation and disposition by exploring avenues of building tangential counter-digital infrastructures from the digital discards for non-discriminatory access and impact.

Tangential Counter-Digital Infrastructures: Dump Yards as Attachment Sites

> "We must ask, 'Who cares?' 'What for?', 'Why do 'we' care?', and mostly, 'How we care?'"
>
> (de la Bellacasa 2011: 96)

These questions by de la Bellacasa interweave with how waste collectors across India generate solidarities among themselves to resist the impact of digital hierarchies. The sites for dumping digital waste in India often function as "attachment sites" for the waste collectors to "tie sticky knots" (Haraway 2008: 287) among themselves and resist the exclusionary designs of digital infrastructures in urban India in general and in Kolkata in particular, by using waste as a tool to create tangential counter-digital infrastructures through recycling. The attachment sites enable these vulnerable communities to "search for more livable 'other worlds' (autre-mondalisations) inside earthly complexity than one could have ever have imagined" (Haraway 2008: 41) and create opportunities for "world making" through "touch, regard, looking back, [and] becoming with" (Haraway 2008: 36).

For instance, Hemanti Devi, a Dalit woman resident of Mainpura slum in Patna's Nehru Nagar and a waste collector by profession mostly collecting dry waste and discarded objects from different localities, started an e-waste processing plant in 2018 so that the processes of cleaning digital waste can be carried out systematically and hygienically by the cleaners. The e-waste processing plant also helped Hemanti and fellow waste collectors transform digital waste into low-cost electronic objects like mobile phones and computer systems for market use and sale (Mehra 2018). These initiatives have been digitally and financially empowering them.

A Muslim Dalit man named Mansoor took a similar transformative initiative in 2015. Mansoor started his career as a waste picker in the Jayanagar area of Bangalore, where he mostly collected discarded digital objects. Gradually, along with his fellow collectors, he led the foundation of the Dry Waste Collection Centre, which collects "10–12 tons of dry waste every month and sorts the same in 72 different categories before it goes for recycling" (Pareek 2015). Besides recycling the dry waste, the discarded digital objects are segregated, converted into low-cost usable digital objects, and made available for everyday use. Along with a local NGO named Hasiru Dala, Mansoor has also founded the Clean City Recyclers Association (CCRA), which is digitally and socioeconomically empowering the cleaners by developing "android apps" and providing smartphones for managing daily cleaning operations (Pareek 2015).

These initiatives have allowed people like Hemanti, Mansoor, and other waste collectors across urban India to interrogate and dismantle the hierarchies in producing and accessing digital infrastructures and build "sibling solidarities" that are carved out "not through identification as sameness but rather through feeling of relation/relatedness with another" (Yengde 2023: 23). Apart from generating access to low-cost digital infrastructures, the values of sibling solidarities are also reflected in the ways in which the waste collectors are giving access to digital infrastructures to their colleagues and sensitizing them about putting the gadgets and platforms to positive use. However, as discussed in the concluding section, the practices of solidarities and relatedness are somewhat irregular and scattered and insufficient to prevent India from being pushed towards a digitally garbocratic future.

Digital Empires and Garbocratic Futures

Concerning the discussions on the toxicities of digital discards, exclusionary intentions of digital infrastructures, and socioeconomic marginalization in this chapter, India appears to be moving towards a future dictated and governed by digital empires with garbocratic principles. From the perspective of this chapter, the phenomenon of garbocracy functions in the context of digital waste, and it outlines how the volumes and patterns of disposing of digital waste across different urban areas in India unearth the sociocultural discrepancies in the development of digital infrastructures.

If these hierarchies and fractions of garbocracy are not effectively countered, the rise of digital empires in usable and non-usable electronic objects is inevitable, where the intra-human existential hierarchies will no longer be regulated by humans but by digital gadgets. In the name of improving and modernizing digital infrastructures, the illogicalities of digitization, such as celebrating improperly disposed digital waste and socioeconomic differences as markers of digital infrastructural progress, have started invading and controlling the ethical, economic, and cultural aspects of human existence.

The purpose of this critique is not to disregard the importance of digitization in contemporary times but to generate an urgent awareness against the onset of a "distorted, terrifying" and "geoengineered" future (Holloway 2022: 60), eroded of the fundamental human values of caring and sharing. If the increasing seductions of hyper-digitalism produced by urban India's digital infrastructures are not identified and controlled, specific human communities will continue to transform other human communities into waste and erase them in the name of "greater aesthetic ... good" (Doron and Jeffrey 2018: 43). This "mutually assured vulnerability" (Bauman 2004: 7) has already generated a state of "derelationalization" (Gordon 2022: 1579) between humans, which soon could be unrecoverable.

Works Cited

Armiero, M. (2021). *Wasteocene: Stories from the Global Dump*. Cambridge: Cambridge University Press.

Ammons A. R. (2003). *Garbage: A Poem*. New York: W.W. Norton & Company.

Arya, S. & Kumar, S. (2020). "E-Waste in India at a Glance: Current Trends, Regulations, Challenges and Management Strategies", *Journal of Cleaner Production*, 271. https://doi.org/10.1016/j.jclepro.2020.122707.

Bauman Z. (2004). *Wasted Lives: Modernity and Its Outcastes*. Cambridge: Polity Press.

de la Bellacasa M. P. (2011). "Matters of Care in Technoscience: Assembling Neglected Things", *Social Studies of Science*, 41(1), 85–106.

Dlamini, J. (2009). *Native Nostalgia*. Cape Town & Johannesburg: Jacana.

Doron, A. & Jeffrey, R. (2018). *Waste of a Nation: Garbage and Growth in India*. Cambridge: Cambridge University Press.

Feenan, D. (2007). "Understanding Disadvantage Partly through an Epistemology of Ignorance", *Social & Legal Studies*, 16(4), 509–531.

Ghosh, R. (2021). "The Plastic Turn", *Diacritics*, 49(1), 64–85.

Gordon, L. R. (2022). "Fanon on Cadavers, Madness and the Damned", *European Journal of Philosophy*, 30 (4), 1577–1582.

Gray-Cosgrove, C., Liboiron, M., & Lepawsky, J. (2015). "The Challenges of Temporality to Depollution & Remediation", *S.A.P.I.EN.S. Surveys and Perspectives Integrating Environment and Society*, 8 (1). https://sapiens.revues.org/1740.

Hage, G. (2009). "Waiting Out the Crisis: On Stuckedess and Governmentality". In G. Hage (ed.), *Waiting*, pp. 97–106. Melbourne: Melbourne University Press.
Haraway, D. J. (2008). *When Species Meet*. Minnesota: University of Minnesota Press.
Holloway, T. (2022). *How to Live at the End of the World: Theory, Art and Politics for the Anthropocene*. Stanford: Stanford University Press.
Hullemane, S. (2023). "Creating a Sustainable Digital Infrastructure Platform for India's Rural Education", *The Economic Times*, <https://economictimes.indiatimes.com/news/india/creating-a-sustainable-digital-infrastructure-platform-for-indias-rural-education/articleshow/102272161.cms?from=mdr>, accessed on 31 January 2024.
Humes, E. (2012). *Garbology: Our Dirty Love Affairs with Trash*. New York: Avery.
Jadhav, R. (2022). "Around 78% of India's e-Waste Is Not Being Collected or Disposed by the Government", *The Hindu*, <https://www.thehindubusinessline.com/data-stories/data-focus/around-78-of-indias-e-waste-is-not-being-collected-or-disposed-by-the-government/article65406820.ece>, accessed on 29 June 2023.
Jayapradha, A. (2015). "Scenario of E-Waste in India and Application of New Recycling Approaches For E-Waste Management", *Journal of Chemical and Pharmaceutical Research*, 7 (3), 232–238.
Latour, B. (2004). "Why has Critique Run out of Stream? From Matters of Fact to Matters of Concern", *Critical Inquiry*, 30(2), 225–248.
Latour, B. (2005). *Reassembling the Social: An Introduction to Actor Network Theory*. Oxford: Oxford University Press.
Lepawsky, J. (2018). *Reassembling Rubbish: Worlding Electronic Waste*. Cambridge & London: The MIT Press.
Luton. L. S. (1996). *The Politics of Garbage: A Community Perspective on Solid Waste Policy*. Ottawa: University of Pittsburg Press.
MacBride, S. (2012). *Recycling Reconsidered: The Present Failure and Future Promise of Environmental Action in the United States*. Cambridge & London: The MIT Press.
Mayaram, A. & Agarwal, B. (2022). "Retooling India's Digital Infrastructure to Truly Help the Poorest", *The Wire*, <https://thewire.in/economy/retooling-indias-digital-infrastructure-to-truly-help-the-poorest>, accessed on 28 May 2023.
Mehra, P. (2018). "From Rag-Pickers to Manufacturers: Tapping into Operation Clean Up", *The Hindu*, <https://www.thehindubusinessline.com/specials/clean-tech/e-waste-collection-and-recycling-how-organisation-is-necessary/article22597548.ece>, accessed on 30 March 2020.
Pareek, S. (2015). "How This Waste Collector in Bengaluru Is Making Entrepreneurs Out of Ragpickers", *The Better India*, <https://www.thebetterindia.com/18324/how-this-waste-collector-bengaluru-making-entrepreneurs-out-of-ragpickers/>, accessed on 15 February 2016.

Pendharkar, V. (2018). "Indian Cities Stare at a Mountain of e-Waste, with Little Idea of How to Manage", *Citizen Matters*, <https://citizenmatters.in/indian-cities-stare-at-a-mountain-of-e-waste-with-little-idea-of-how-to-manage-it-6874>, accessed on 11 June 2018.

Pinto, V. N. (2008). "E-waste Hazard: The Impending Challenge", *Indian Journal of Occupational and Environmental Medicine*, 12(2), 65–70.

Strach, P. & Sullivan, K. S. (2023). *The Politics of Trash: How Governments Used Corruption to Clean Cities, 1890–1929*. Cornell: Cornell University Press.

Teo, M. M. M. & Loosemore, M. (2010). "A Theory of Waste Behaviour in the Construction Industry", *Construction Management and Economics*, 19(7), 741–751.

Yengde, S. (2023). "Dalit in Black America: Race, Caste, and the Making of Dalit-Black Archives", *Public Culture*, 35 (1), 21–41.

CHAPTER 8

Conclusion: Towards Counter-Garbocratic Futures

Physical and ideological garbage has now become a tentacular garbocratic empire, suffocating not only the present but also the future. It has a "decisive advantage over its predecessors, the popularity of its present, and the messianic promise of a future" (Ghosh 2022: 15). A garbocratic empire embodies "pure spontaneity without obvious intent" (Pelkonen 2015: 4). It is turbulent, "conflictual and negotiatory" (Ghosh 150). It is a living entity with which "we simulate, then transplant every facet of reality, converting all the varied elements of the planet into one common emulsion" (Bok 2016: 25). It is ideologically, psychologically and intellectually paralytic and destructive, and is "multiplying exponentially" across India and the world (Craps 2023: 71). Within diverse social, political, cultural, and economic contexts, the empire may have evolved locally in India but has been influencing and regulating the planet in diverse emotional and ideological ways. This is why, despite the widely exposed socioeconomic gaps, India is celebrated as one of the world's economic superpowers, based on the rise in the number of billionaires in the country; despite the dictatorial, divisive, aggressive, and militaristic governing policies, the BJP government is acknowledged as one of the most idealistic and idolistic governments worldwide; despite the rise in queerphobia, Islamophobia, and anti-Dalit violence, India continues to be appreciated as one of the most inclusive and diverse countries; and despite the reckless destruction of the natural environment and a tremendous rise in environmental catastrophes, India is understood as a country with one of the world's most balanced economy.

These falsified narratives are maintained not only by agencies within India but also by other agencies. They are collectively upheld by varied right-wing and pro-right-wing agencies safely and stealthily housed within the neocolonial and neoliberal educational institutions, political organizations, and public spaces of the Global North. For example, let us analyze how Infinity Foundation Official, a non-profit organization based in Princeton, propagates the toxic principles of Hindutva across the US and the world in the name of reviving and universalizing India's pre-colonial cultural and traditional past. The organization was founded by Rajiv Malhotra, an Indian-born American Hindutva ideologue who, through diverse lectures, discussions, and research projects, tries to prove the superiority of Hindu and Hindu-centric cultures and societies over other cultures, institutions, and spiritualities. His YouTube channel's content may appear to be mere documentation of Hindu knowledge systems. However, a deeper analysis reveals that he serves as an international mouthpiece for the BJP. His ideas, ideologies, and propaganda work to create a planetary network of infectious, diseased, and deceptive right-wing neoliberal individuals and institutions who blindly volunteer the agendas and propaganda of right-wing intellectuality and governmentality across the world. Anyone who questions his intentions is violently rejected and silenced.

For instance, in 2018, Rajiv delivered a lecture on "Decolonizing Academics" at Oxford University. In the lecture, he argued that Ivy League, elitist, white-centric, and heteronormative institutions like Oxford University and others continue to colonize knowledge systems through their Eurocentric and racist curricula, pedagogies, and infrastructures, and that "answering back" (Malhotra 2018) to them against their corrupt geopolitical designs is still considered inappropriate and unethical. However, the narratives he used to justify his arguments were subtly directed towards the Hindutva vision of the BJP government and how India's decolonial revolution began with them. When a European person from the audience critiqued that his lecture was less intended towards decolonizing the education system and more intended towards celebrating the garbageous, hollow, and propagandist knowledge-making systems of India, the critique was immediately silenced, shunned off, and forced out of the lecture hall by the speaker and a few pro-Hindutva Indian students, who were close associates of Rajiv.

This was not the first time he and his close associates adopted a rejectionist approach towards individuals questioning his arguments. In 2015, during a roundtable discussion at Jawaharlal Nehru University on caste hierarchies, representation of Indian cultures abroad, and the position of Dalits in India and abroad, Udit Raj, then a Member of Parliament for BJP (2014–2019), was critiqued and expelled from the room by students and faculty members for arguing in favour of Dalits who have settled abroad and for seeking foreign investments for Dalits in India, which Rajiv and many in the audience regarded as a colonial attitude (Infinity Foundation Official 2015). When Udit argued that the high-caste elites were undertaking similar investment initiatives to enhance financial flow in India, efforts were immediately made to silence him and redirect the arguments. What is also interesting is that, despite being part of the BJP, Udit was silenced because he raised a sensitive and reveling topic for the party. Furthermore, a careful listen to the discussions reveals that Rajiv's supporters made every effort to ceremonialize the fractured, superficial, and garbocratic governing methodologies of the BJP. His YouTube channel also features discussions with and about controversial Hindu spiritual leaders like Baba Ramdev, Swami Nithyananda,[1] and others regarding their social contributions, methods to condemn Hinduphobia, and the necessity of building a Hindu global order. Both Ramdev and Nithyananda have been legally sued for promoting fake medicinal, scientific, cultural, economic, and political narratives cluttered with ignorance, denial, divisiveness, and hatred.

The tendency to revive the colonially structured 'divide, silence, and rule' policies of the Europeans in India can also be observed in the case of the Kuki-Meitei ethnic conflicts in Manipur in 2023. Though the BJP-ruled state government brushed aside the ethnic motivations behind this conflict, blaming it on drug lords and antisocial activists entering the state, the sequence of events clearly shows that the conflicts and pogroms were ethnically motivated. The Kuki and Meitei people are ethnic residents of Manipur, and their conflictual existence dates back to the 1980s, when

---

[1] The discussions can be accessed through the links: <https://www.youtube.com/watch?v=CTZavpY5SFs&t=12s> and <https://www.youtube.com/watch?v=dvcJI5yAd6M&t=1852s>.

the then Chief Minister of Manipur demanded ST (Scheduled Tribe) status for the Hindu-majority Meitei community in the State Legislative Assembly. However, besides resistance from the Kukis, the proposal was opposed by some Meiteis as well. In May 2023, tensions were reignited when the Manipur High Court recommended that the "Bharatiya Janata Party (BJP) government in the state to include Meiteis in the Scheduled Tribe category" (Rathore 2023; also see Deka 2023). The recommendation for the Hindu-majority Meiteis was underpinned by specific political intentions and aligns with the Hindutva project of cultural, communal, caste, and religious divisions deployed across all other BJP-ruled states in India, akin to the ethnic and racial divisions promoted during British rule. Historically, since the European colonial era, Indigenous communities in Northeast India have been socially, culturally, economically, and racially treated as unwanted objects or waste, to be rejected, mashed, moulded, and discarded at any point.

After India's judicial independence, the neo-colonization of northeastern India continued with the launch of the Armed Forces Special Powers Act (AFSPA) in 1958, which granted special powers to the Indian armed forces, the state, and paramilitary forces. As a result, the communities are habitually subjected to physical and psychological harassment, residential displacements, and rape in the name of law, order, security, and discipline. With the emergence of BJP rule in Manipur and many other parts of the Northeast, a deep communal angle has been added to the existing sociopolitical cleavages.

However, it is essential to note that the garbageous present and the impending garbocratic future have not gone unchallenged. Multiple social, cultural, economic, and political initiatives are being taken to counter the crises discussed in this book. In the context of decapitalizing political-ecological coexistence and co-functioning in India, let us examine some of the environmental and economic initiatives being taken to nurture a paradigm of development and growth that is disentangled from Euro-capitalocentric parameters of development and does not demean or demonize selective human and more-than-human communities. They include:

a. *Give Me Trees*: Give Me Trees is a pan-Indian environmental conservation project started by Swami Prem Parivartan, popularly known as Peepal Baba. This project aims to generate "community education, awareness, and volunteer mobilization campaigns" (Peepal Baba 2024) to save India from deforestation and environmental destruction. Along with several volunteers, Prem Parivartan has planted 45 million trees along roadsides, highways, polluted rivers, and unclaimed barren lands across the country to revive the natural green covers on the one hand and to prevent unsystematic dumping of garbage in public areas on the other. In addition to planting "goolar, … pilkhan and also banyan and neem", he has planted 20 lakhs of peepal trees "for its medicinal qualities and because it supports a whole eco-system" (Khandekar 2013).

Besides planting trees, through varied social media channels, he teaches household techniques for planting, flowering, watering, mulching, and conserving plants and soils so that individuals can undertake low-cost and environmentally sustainable greening initiatives in their localities. He also teaches how to use household waste like wasted foods, teabags, unused water, and others to prepare biofertilizers for the plants' physical and mental well-being (Peepal Baba 2021). These initiatives are not just taken for environmental conservation but to project an alternative possibility in which developmental projects can be fostered without incessant accumulation of garbage, rash construction of residential complexes and shopping malls, abusive displacements of vulnerable communities, and uncontrollable building of factories and industries.

b. *Garbage Café in India*: In addition to Give Me Trees' ecological revival initiatives, efforts are also being made to reduce garbage consumption and disposal in innovative ways, such as constructing garbage cafés across India. The local municipal corporation of Ambikapur, Jharkhand, took up this unique initiative in 2019. This socioeconomic initiative is intended for the ragpickers who receive free meals in exchange for 1 kg of plastic garbage. Prior to the establishment of the garbage café, Ambikapur had already gained prominence as the second cleanest city in India after Bhopal. As part of this scheme, "the collected garbage will be sold at the Solid-Liquid Resources Management (SLRM) Centre from where a coupon will be issued. This coupon will allow people to avail breakfast or lunch at the canteen at Ambikapur bus stand" (Kaur 2019). The plastic waste is sent to a "corporation-run recycling plant that turns it into granules for sale" (Kaur 2019). This initiative has generated multiple eco-sustainable modes of living, such as using plastic granules to construct roads, making Ambikapur dustbin-free, and generating 12 lakh rupees per month by selling recycled paper and plastic granules. Ambikapur exemplifies how sustainability and profitmaking strategies can be balanced without compromising hygiene, psychological peace, aesthetic visions, and cohabitation. There was a time when Ambikapur was a dump site; today, it boasts a "14-acre botanical garden" (Kaur 2019).

Garbage Café in Ambikapur became a success primarily because of the collective consciousness of the residents, who realized the damage that garbage can physically, psychologically, socially, and culturally cause to individuals and communities. Apart from a compromised state of hygiene, when Ambikapur was a dumping ground (before 2015), the locality was also experiencing severe economic losses due to untraversable conditions (All India 2019). Except for the residents, the shops had no other customers. The situation worsened further when residents started moving out to different localities. These crises triggered the remaining residents to bring paradigmatic changes, which is how the garbage café project was born.

c. *Plasticulture*: The agricultural phenomenon of plasticulture is another creative method through which plastic waste can be reduced and channelled towards a more sustainable direction by blending plastic with agricultural growth. The use of plasticulture in India was streamlined with the establishment of the National Committee on Plasticulture Applications in Horticulture (NCPA) in 2001. The NCPA proposed that, to reduce plastic waste in India, everyday plastic waste could be put to sustainable use through "seeds packaging, planting, propagation, mulching, irrigation, harvesting, fruit packing and preservation" (Sharma and Chandra 2019: 44). In addition to these aspects, plasticulture also assists in "soil fumigation/soil solarization," "propagation and nursery management," "protected cultivation," "drip irrigation," and "pest management" (Sharma and Chandra 44–45) and in various other agricultural and commercial areas. However, it is crucial to mention that plasticulture does not come without limitations. When NCPA launched this initiative, the main intention was to reduce plastic consumption, which eventually did not work out. On the contrary, these initiatives were often used to normalize the country's plastic use.

Despite these challenges, the use of plasticulture demonstrates what Mieke Bal describes as " 'symbiosis' as an 'inter-ship' " (2023: 79). According to Bal, unlike the "facile 'trans-', which means nothing but traversing without, precisely, engaging …" 'inter-' means "between in relationship with" (Bal 79). The symbiosis of plastic and agriculture potentially draws a similar picture. In India, the practice of plasticulture is still in its nascent stage. It may not be an ideal solution to reduce plastic garbage in India, but it does make us realize that "the only thing that does not seem to matter anymore is matter," and beyond matter, a lot of things matter, like "Language matters. Discourse matters. Culture matters" (Barad 2003: 801). Plasticulture has reduced the "stilling, halting, [and] fixing" (Bal 81) of personal and public existential spaces by plastics and other plastic-associated wastes. Plasticulture is not only ecologically transforming the agricultural system in India but is also teaching humans to build a more "compost" (Haraway 2016a: 97) society, underpinned by the rituals of human, other-than-human, bodily, and technological coexistence (Forster 2022). In the present-day era of garbage, garbaging, and garboglomerates, habitual existence is shaped by a "top-down, master-planned vision" that is centred "around the needs of the suppliers rather than the needs of the citizens" (Calzada 2021: 18;

also see Halegoua 2020). Initiatives, like Give Me Tree, Garbage Café, and plasticulture technologies, have the potential to obstruct garbocratic developments.

d. *Solar homes made with junk*: Another initiative that can be cited as a prominent example to prevent a garbocratic future for India is making solar homes with junk. In 2019, a Chennai-based architect named Arun Prabhu constructed "a house on top of an auto-rickshaw that could be a long-term solution to illegal encroachments in metro cities" (Karelia 2021). Developmental projects in metro cities are fuelled by "technological utopianism, neoliberalism, security and surveillance, and digital citizenship" (Karpouzou 2023: 102), and all of these aspects structurally ignore the fundamental crises of hygiene, health, residency, and the natural environment. Due to these challenges, illegal encroachments are a significant concern. Arun has developed a low-cost solution: a house on top of an auto-rickshaw to find a possible solution. The house is called 'Solo 0.1' and has "a toilet, foyer, terrace, living room space, kitchen, and other essential spaces" (Karelia 2021). According to Arun, this home will not only address the residential and pollution challenges in cities, but will also assist "solo travelers, vendors, artists, homeless and construction workers" (Karelia 2021) in finding quick residential solutions.

Auto-rickshaw homes may also prove beneficial during evacuations and natural disasters. Arun's discovery demonstrates that integrating environmental, digital, technological, and human systems can occur without destruction, rejections, and erasures. All these systems can be respectfully, mutually, and warmly embraced through the social values of ecological intercorporeality, where every form of living body comes a bit closer without pushing each other out. Within an intercorporeal space of "social cognition" (Tanaka 2015: 455), psychologically, symbiotically, and semiotically, every living being is tied to one another. Arun's auto-rickshaw home promises such a closely related and deeply entangled existential space of accessibility and togetherness that is horizontally structured – far from the egocentric, violent, and exclusionary designs of garbocratic existence.

The use of biogas to light up the entire Varadharajapuram village in Tamil Nadu (Sivapriyan 2021) can be cited as another crucial counter-strategy to escape a garbageous mode of thinking, acting and living in the present and the future. Nonetheless, these initiatives are insufficient to tackle the garbocratic "ecoprecarity" (Nayar 2019: 7) and to bulldoze the enormous mountains of garbageous existence accumulated so far. Despite these initiatives, the presence and persistence of plastics, lead, zinc, and other toxic materials within humans and non-humans are rapidly increasing in the forms of microplastics, lead powders, and zinc powders, infecting and choking countless living beings to death. Microplastics are especially "like a virus" (Robi, cited in Nasution & Laksono 2021). They are "invisible to the naked eye, but the threat is serious" (Robi cited in Nasution & Laksono 2021). Microplastics and other pollutants affect the liver, testicular systems, and renal systems and may cause cancer and other fatal diseases. The production of these materials is associated with the garbageous socio-political-economic-commercial-industrial-cultural-historical-communal-imperial nexus of planetary destruction. Therefore, more collective involvement and self-realization are required to counter them.

Self-Resurrection

One of the possible ways in which collective self-realization and resistance can be further stimulated is self-resurrection. The phenomenon of self-resurrection can be implemented in the forms of self-interrogation and self-transformation. It is commonly observed that while addressing any form of public challenge, the primary focus remains on changing governments, policy transformations, and enhancing scrutiny measures through public surveillance and censorship. Though many conversations surround the 'notion' of self-consciousness and the necessity of adopting it as a chronic existential measure, hardly any importance is given to putting the 'notion' into 'practice'. To explain further, self-consciousness and self-transformation are commonly focused perspectives during public and official discussions. They have been ritualistically and mechanically highlighted so many times across multiple social, cultural, political, public, and personal spaces that they have been diluted within the psyches across generations. They are only parroted and repeated during discussions.

The process of self-resurrection through self-realization needs to be generated right from childhood, where children, instead of being imposed with predetermined knowledge systems, should be taught how to think critically and freely. Due to a significant lack of critical thinking, when voting machines are rigged, vote banks are manipulated, and people are openly beaten up and murdered in public for expressing their opinions, rights, and duties fearlessly, then out of fear, the public remains silent or justifies the violence through conjectured parameters of security and discipline. Developing the capability of free thinking is not restricted to blatant text-centric and degree-centric learning, but expands across diverse modes and patterns of lived experiences, from which children are often guarded in the name of protection, love, and care. As a result, when children grow up, a lot of the time, the only tool that they are left with to interpret the societies around them is a prison of pre-established ideologies, lenses, and virtues in which they have been caged, manipulated, tortured and silenced since they were born. Any form of a functional, intellectual, and behavioural pattern that exists outside this prison is immediately downplayed as unethical and anti-disciplinary. The well-executed and ideologically circumcised systems of governing selective bodies, minds, emotions, places, and spaces must be habitually interrogated in diverse personal and collective contexts in self-reflective, self-critical, and unprejudiced ways.

Ecological Resurrection

A possibly successful self-resurrection may lead to ecological resurrection. Again, engaging with ideas and policies about environmental sustainability is not new. Since humans' systemic and structural ruination of the natural environment, endless discussions and arguments have occurred to prevent the regularized damage to nature and revive the natural surroundings. However, what continues to be unrealized is that the revival of nature cannot be simultaneously practised along with the destruction of nature. Capitalistic phenomena like eco-capitalism, green capitalism, and green industrialization have normalized a culture of annihilative revivalism, which continues to celebrate reckless profitmaking agendas by curating fake ecological consciousness. The phenomenological exercise of ecological resurrection exposes this fakeness. It provokes us to figure out genuine pathways of respecting and reviving the natural environment around us through "personal commitment and pluralistic elaborations" (Ferrando 2023: 10).

Ecological resurrection is not just about co-learning and cohabiting with the natural environment, but is also an awareness that we must keep on "re-evaluating ourselves as planetary beings ... in awareness and response-ability" (Ferrando 10). Ignorance of this realization will only "endanger our survival" (Ferrando 10) and give birth to garbocratic futures. We are already choked enough with garbageous aspects like physical waste, environmental disasters, warfare, communal violence, sociopolitical dictatorship, and economic disintegration. The lack of ecological resurrection or remaining associated with pseudo-revival strategies will only worsen the existential conditions of humans and other-than-humans.

Cultural Resurrection

The necessity of ecological resurrection also draws our attention towards the other-more-than-human cultures and politics of "shared precarities" (Karpouzou and Zampaki 2023: 11) and "unchosen proximities" (Cohen 2015: 107) that are underpinned by a plexus of "material flows, exchanges, and interactions of substances, habitats, places, and environments" (Alaimo 2011: 281). While outlining governing policies in India and other parts of the world, many conversations occur about building collective environmental consciousness through constructing Urban Green Spaces (UGS) like botanical gardens, city forests, eco-parks, and others. Still, hardly any policies have been developed to leave natural green spaces undisturbed and untouched. The necessary developmental projects of constructing industries, factories, residences, roads, railways, and shopping complexes can be undertaken by keeping the natural environment at the centre.

The cultures and rituals of practising development and modernity need to be critically revisited, re-interrogated, and expanded beyond the paradigm of anthropocentrism. Once "we underline the human not as one but as many, some may emphasize that other notions and practices – such as symbiosis, affinity, and so on – are as fundamental as the category of alterity [...]" (Ferrando 2019: 70). The phenomenon of cultural resurrection provokes us to curate co-existential conditions "in which the decentering of the human, its imbrication in technical, medical, informatics, and economic networks are increasingly impossible to ignore" (Wolfe 2010: xv).

For example, how an anti-garbage and anti-garbocratic project in Bangalore named Hasiru Dala evolved through exercises of self, cultural, and ecological resurrections. Hasiru Dala means "green force," and it is a "member-based organization of informal waste workers" (Hasiru Dala 2015). The organization was started in 2011 to provide waste sorters and waste pickers with "opportunities that will help them fight discrimination and gain recognition for the contribution they make" (Hasiru Dala 2015). This organization gives entrepreneurial empowerment to the waste pickers by providing them with official ID cards, bank accounts, scholarships, and education loans. These initiatives have given the waste collection job in Bangalore a respectful identity. The organization also assists the city's Dry Waste Collection Centres (DWCC) in collecting waste and scraps and processing them in organic ways. The initiatives of Hasiru Dala ensure the city's cleanliness and contribute to demolishing the caste, class, social, cultural, and economic prejudices associated with the waste pickers and the cleaners. Initiatives like this can effectively counter the impending garbocratic future by "interrogating and erasing the boundary that has been assumed" to set specific privileged human communities apart "from the rest of the living community" (Westling 2006: 30). These eco-philosophies of

Conclusion: Towards Counter-Garbocratic Futures

"multiple belongings" (Karpouzou and Zampaki 2023: 13) have the potential to generate counter-garbocratic futures of multirooted symbiocracies and archipelagic solidarities.

Multirooted Symbiocracies and Archipelagic Solidarities: Towards Counter-Garbocratic Futures

Having discussed the different forms of garbageous and garbocratic existence in the present times and the possible counter-garbocratic resistances, it is essential to wrap up this book by exploring the shifts from garbocracies towards multirooted symbiocracies. The phenomenon of multirooted symbiocracies emerges from a state of "symbiogenesis" (Haraway 2016a: 218; also see Haraway 2015), where different living entities are cobbled together "to make something new in the biological, rather than digital or some other mode" (Karpouzou 2023: 114). The social, cultural, and intellectual patterns of existence of every human and other-than-human species are historically and genetically multirooted and transborder. However, colonial and capitalistic knowledge systems to manufacture and maintain fractions, hierarchies, and crises never accepted this reality, eventually resulting in a "waste apocalypse" (Ghosh 2022: 115) – an apocalyptic state of wasteful meaning-making, thinking and doing.

As discussed in this book, the regularization of waste apocalypse reaches a climactic state when the rhizomatic bonds between every living species are torn away to make way for violent and abusive segregations and fragmentations as normative living. This existential state of what Patricia Yaeger interprets as the "apotheosis of trash" (Yaeger 2008) blocks all the possible corridors of "interspecies communication" (Karpouzou 2023: 117). The possibility of communication with different species enhances the quality of co-habitation and collaborations between humans and teaches us how to "live in synergy with our plant and animal friends" (White, quoted in Karpouzou 117). Multirooted symbiocracies serve as the connecting point of intersemiotic articulations between humans and humans and humans and other-than-humans. Multirooted symbiocracies are a potential "alternative to authoritarian-based structures, [...] that could better regulate interaction between individual Space settlers, their counterparts on Earth, as well as between multiple settlements in the future" (Lockard 2012: 1).

Apart from multirooted symbiocracies, co-building archipelagic solidarities could be another effective way of countering garbocratic futures. Like a group of islands co-existing without trying to assimilate, annihilate, and appropriate each other, the phenomenon of archipelagic solidarities invites humans and other-than-humans to establish "one's being in relation to others [...] not through identification as sameness, but rather through a feeling of relation/relatedness with another" (Yengde 2023: 22–23). Archipelagic solidarities can also generate tangential pathways of existence by splintering and re-constellating (Kabir 2022: 1) the garboglomerates as symbioglomerates. Symbioglomerates can be understood as "subterranean networks" (Burchardt and Leinius 2022: 5) of "myriad intra-active entities-in-assemblages – including the more-than-human, other-than-human, inhuman, and human-as-humus" (Haraway 2015: 160). Multirooted symbiocracies and archipelagic solidarities may not be the ultimate solutions to resolve the garbageous patterns of physiological, ideological, and ecological crises discussed in this book regarding politics, cultures, societies, and the natural environment. Still, they offer us concrete possibilities of co-curating alternate futures of "tentacular thinking" (Haraway 2016b), knitted with the threads of "sympoiesis, symbiosis, symbiogenesis, development, [and] webbed ecologies ..." (Haraway 2016b). So, to "make-with – become-with" (Haraway 2015: 161) multiple species, ideologies, psyches, and emotions. Let the sympoietic co-labouring begin.

Works Cited

Alaimo, S. (2011). "New Materialism, Old Humanisms, or Following the Submersible", *NORA – Nordic Journal of Feminist and Gender Research*, 19 (4), 280–284.

All India. (2019). "'Garbage Café' in Chhattisgarh to Offer Free Food in Exchange of Plastic", *NDTV*, <https://www.ndtv.com/india-news/garbage-cafe-in-chhattisgarh-to-offer-food-in-exchange-for-plastic-2074828>, accessed on 31 January 2023.

Bal, M. (2023). "How to Say It? Symbiosis as Inter-Ship". In P. Karpouzou & N. Zampaki (eds.), *Symbiotic Posthumanist Ecologies in Western Literature, Philosophy and Art: Towards Theory and Practice*, pp. 79–98. Berlin: Peter Lang.

Barad, K. (2001). "Posthumanist Performativity: Toward an Understanding of How Matter Comes to Matter", *Signs*, 28 (3), 801–831.

Bok, C. (2016). "Virtually Nontoxic", *English Studies in Canada*, 42 (3–4), 25–26.

Burchardt, H. -J. & Leinius, J. (2022). "Of Archipelagic Connections and Postcolonial Divides". In H. -J. Burchardt & J. Leinius (eds.), *(Post-)Colonial Archipelagoes: Comparing the Legacies of Spanish Colonialism in Cuba, Puerto Rico, and the Philippines*, pp. 3–19. Ann Arbor: University of Michigan Press.

Calzada, I. (2021). *Smart City Citizenship*. Amsterdam: Elsevier.

Cohen, J. J. (2015). "The Sea Above". In J. Cohen & L. Duckert (eds.), *Elemental Ecocriticism: Thinking with Air, Water, and Fire*, pp. 105–133. Minneapolis: Minnesota University Press.

Craps, S. (2023). "Ecological Mourning: Living with Loss in the Anthropocene". In B. A. Kaplan (ed.), *Critical Memory Studies: New Approaches*, pp. 69–77. London: Bloomsbury.

Deka, K. (2023). "Drugs, Land Rights, Tribal Identity and Illegal Immigration – Why Manipur Is Burning". *India Today NE*, <https://www.indiatodayne.in/manipur/story/drugs-land-rights-tribal-identity-and-illegal-immigration-why-manipur-is-burning-553788-2023-05-05>, accessed on 31 December 2023.

Ferrando, F. (2019). *Philosophical Posthumanism*. New York & London: Bloomsbury.

Ferrando, F. (2023). "'We Are the Earth': Posthumanist Realizations in the Era of the Anthropocene". In P. Karpouzou & N. Zampaki (eds.), *Symbiotic Posthumanist Ecologies in Western Literature, Philosophy and Art: Towards Theory and Practice*, pp. 9–10. Berlin: Peter Lang.

Forster, Y. (2022). "Rituals of Coexistence: Bodies and Technology during Pandemics", *Interlitteraria*, 27 (1), 84–98.

Ghosh, R. (2022). *The Plastic Turn*. Ithaca: Cornell University Press.

Halegoua, G. R. (2020). *Smart Cities*. Cambridge: MIT Press.

Haraway, D. (2015). "Anthropocene, Capitalocene, Plantationocene, Chthulucene: Making Kin", *Environmental Humanities*, 6 (1), 159–165.

Haraway, D. (2016a). *Staying with the Trouble: Making Kin in the Chthulucene*. Durham: Duke University Press.

Haraway, D. (2016b). "Tentacular Thinking: Anthropocene, Capitalocene, Chthulucene", *e-flux Journal*, 75, <https://www.e-flux.com/journal/75/67125/tentacular-thinking-anthropocene-capitalocene-chthulucene/#:~:text=Specifically%2C%20unlike%20either%20the%20Anthropocene,sky%20has%20not%20fallen%20%E2%80%94%20yet>, accessed on 24 October 2023.

Hasiru Dala. (2015). "Waste Pickers to Robust Entrepreneurs: Creating Stories of Change", <https://hasirudala.in/wp-content/uploads/2020/09/HD_Annual_Report_2015-16-1.pdf>, accessed on 25 August 2021.

Infinity Foundation Official. (2015). "Concluding Discussions Turns Into Shouting Match between Students & Udit Raj", *Infinity Foundation Official*, <https://www.youtube.com/watch?v=AuVaei1oDuo&list=PLGQElwzyJxtxzu8ge3d2ybYROVuhDPYwr&index=4>, 18 March 2022.

Kabir, A. J. (2022). "The Creolizing Turn and Its Archipelagic Directions", *Cambridge Journal of Postcolonial Literary Inquiry*, 10 (1), 90–103.

Karelia, G. (2021). "Watch: 23-YO Builds Low-Cost Solar 'Rickshaw' Home from Scrap at 1/5th the Cost", *The Better India*, <https://www.thebetterindia.com/262406/india-innovation-ecofriendly-solar-scrap-rickshaw-low-cost/>, accessed on 17 November 2021.

Karpouzou, P. (2023). "Symbiotic Citizenship in Posthuman Urban Ecosystems: Smart Biocities in Speculative Fiction", In P. Karpouzou & N. Zampaki (eds.), *Symbiotic Posthumanist Ecologies in Western Literature, Philosophy and Art: Towards Theory and Practice*, edited by, pp. 99–121. Berlin: Peter Lang.

Karpouzou, P. & Zampaki, N.(2023). "Introduction: Towards a Symbiosis of Posthumanism and Environmental Humanities or Paving Narratives for the Symbiocene". In P. Karpouzou & N. Zampaki (eds.), *Symbiotic Posthumanist Ecologies in Western Literature, Philosophy and Art: Towards Theory and Practice*, pp. 11–39. Berlin: Peter Lang.

Kaur, C. (2019). "India's first garbage café to come up in Ambikapur", *Times of India*, <https://timesofindia.indiatimes.com/city/raipur/indias-first-garbage-cafe-to-come-up-in-ambikapur/articleshow/70339167.cms>, accessed on 23 August 2020.

Khandekar, N. (2013). "Several trees across Delhi and counting", *Hindustan Times*, <https://www.hindustantimes.com/delhi-news/several-trees-across-delhi-and-counting/story-6AHmKWM63kaRUdnLaVkzgJ.html>, accessed on 31 January 2019.

Lockard, E. (2012). "'Symbiocracy': The Structuring of New Societies in Space Based on the Principles of Mutualism and Symbiotization", *Aerospace Research Central*, 1–10. <https://doi.org/10.2514/6.2008-7817>.

Malhotra, R. (2018). "My Oxford Lecture on 'Decolonizing Academics.'" *Infinity Foundation Official*, <https://www.youtube.com/watch?v=YVlM_a8lmBI>, accessed on 13 March 2021.

Nasution, R. & Laksono, D. D. (Director). (2021). *Plastic Island* [Film]. Netflix.

Nayar, P. K. (2019). *Ecoprecarity: Vulnerable Lives in Literature and Culture*. New York & London: Routledge.

Peepal Baba. "Life Story." <https://www.peepalbaba.org/life-story/>, accessed on 15 March 2024.

Peepal Baba. "Plants and Mental Well-being", *Peepal Baba*, <https://www.peepalbaba.org/plants-and-mental-well-being/>, accessed on 2 August 2021.

Pelkonen, E. -L. (2015). "Plastic Imagination", <https://forty-five.com/pdfs/45-Pelkonen.pdf>, accessed on 30 May 2022.

Rathore, S. (2023). "Navigating the Kuki-Meitei Conflict in India's Manipur State", *The Diplomat*, <https://thediplomat.com/2023/08/navigating-the-kuki-meitei-conflict-in-indias-manipur-state/>, accessed on 1 January 2024.

Sharma, R. & Chandra, A. K. (2019). "Plasticulture Technology: Plasticulture for Profitable Horticulture", *Agriculture & Food: E-Newsletter*, 1 (11), 44–47.

Sivapriyan, E. T. B. (2021). "Turning obstacle into an opportunity: How a Tamil Nadu village solved its problems", *Deccan Herald*, <https://www.deccanherald.com/india/turning-obstacle-into-an-opportunity-how-a-tamil-nadu-village-solved-its-problem-1005157.html>, accessed on 13 July 2022.

Tanaka, S. (2015). "Intercorporeality as a theory of social cognition", *Theory & Psychology*, 25 (4), 455–472.

Westling, L. (2006). "Literature, the Environment, and the Question of the Posthuman". In by C. Gersdorf and S. Mayer (eds.), *Nature in Literary and Cultural Studies: Transatlantic Conversations on Ecocriticism*, pp. 25–47. Amsterdam and New York: Rodopi.

Wolfe, C. (2010). *What is Posthumanism?* Minneapolis and London: University Minnesota Press.

Yaeger, P. (2008). "The Death of Nature and the Apotheosis of Trash; Or Rubbish Ecology", *PMLA*, 123 (2), 321–339.

Yengde, S. (2023). "Dalit in Black America: Race, Caste, and the Making of Dalit-Black Archives", *Public Culture*, 35 (1), 21–41.

Index

ability xii, 33, 52, 80
academic 60, 62, 66, 67, 79, 83, 84, 87, 108
aesthetic 7, 50, 64, 89, 103, 112
Amendment 69, 77, 81
attitude 4, 8, 10, 14, 25, 32, 34, 37, 38, 51, 63, 64, 66, 67, 69, 70, 84, 96, 98, 109

Bharatiya Janata Party (BJP) ix, xvi, 9, 33, 85, 107, 108, 109, 110
body 4, 14, 16, 22, 25, 40, 62, 69, 114
British 37, 47, 49, 77, 78, 81, 118
Burn 5, 8, 24, 40, 94, 98

caste ix, 6, 8, 9, 10, 12, 14, 22, 25, 26, 27, 32, 37, 39, 40, 49, 51, 81, 96, 109, 110, 118
civilization 11, 20, 37, 40, 61, 77, 88
class 6, 12, 14, 15, 22, 25, 27, 39, 47, 49, 63, 65, 66, 68, 77, 78, 79, 80, 81, 83, 88, 89, 93, 94, 96, 97, 100, 118
coloniality 39, 40, 47, 49, 78, 81, 82, 83, 85, 108, 109, 110, 119
consciousness 3, 13, 20, 50, 54, 64, 66, 69, 112, 115, 117, 118
corruption 79
counter-garbocratic 15, 107, 119
culture 64, 113, 117

Dalit xv, 8, 40, 50, 51, 61, 66, 98, 99, 101, 107
develop xii, xvii, 10, 12, 15, 34, 39, 46, 48, 50, 64, 65, 79, 83, 88, 97, 99, 101, 102, 110, 111, 114, 116, 118, 120

discard xi, 26, 60, 69, 94
Discardscapes 2, 93, 97, 98
dispose 6, 8, 9, 10, 12, 13, 21, 23, 25, 32, 36, 80, 94, 96, 99, 102

ecology 47
economy 47, 70
ecosystem 26
education xv, 33, 39, 47, 59, 60, 62, 63, 65, 67, 70, 80, 81, 82, 83, 108, 111, 118
ego 83, 115
effective 102, 118, 120
egalitarian xii
election 47, 48, 49
empire 11, 13, 19, 21, 23, 25, 27, 107
environment ix, 5, 9, 45, 46, 77, 78, 82, 83, 107, 114, 117, 118, 120

factors 3, 4, 5, 6, 8, 31, 77, 82
fake 33, 36, 46, 49, 51, 81, 99, 109, 117
forest 46, 61, 78, 79, 111, 118
fracture xvi, 8, 39, 45, 88, 109
future 15, 37, 45, 47, 64, 89, 90, 93, 102, 107, 110, 114, 115, 117, 118, 119, 120

Gandhi, Mahatma 33, 37, 38
garbo-being 6, 7, 14, 45, 47, 49, 50, 54, 85
garbo-citizenship 15, 77, 79, 82, 83, 84, 85, 86
garbocratic 15, 37, 49, 61, 64, 78, 80, 90, 93, 102, 107, 109, 110, 114, 115, 117, 118, 119, 120
garbo-curricula 15, 59, 63, 64, 69
garbo-ideologies 84, 15, 41, 45, 48, 50, 54, 85

garbo-imperialism 13, 14, 19, 85
garbo-intellectualism 14, 31
garbo-knowledge 14, 31, 85
garbo-pedagogies 2, 15, 59, 63, 64, 69, 70
garbo-philosophy 36
garbo-power 14, 41, 45, 50, 54, 85

Hindu ix, 4, 8, 10, 17, 23, 24, 26, 33, 34, 37, 38, 39, 43, 44, 48, 49, 51, 55, 68, 79, 81, 82, 84, 104, 108, 109, 110, 112
history 3, 4, 5, 8, 9, 10, 25, 26, 27, 31, 36, 37, 38, 39, 41, 42, 43, 60, 61, 63, 66, 67, 77, 84, 85, 88, 97, 98, 110, 115, 119
humanities 60, 63, 65
hygiene 3, 4, 6, 8, 9, 22, 98, 101, 112, 114

ideology 37, 41, 90
infrastructures viii, 15, 93, 97, 99, 100, 101, 102, 103, 108, 112
imperialism viii, xvii, 13, 14, 19, 21, 23, 25, 27, 73, 85, 112
Indigenous xvi, 34, 39, 41, 46, 61, 66, 78, 110
industrialization 45, 82, 117
intellect 10, 14, 15, 27, 31, 32, 33, 34, 36, 37, 38, 39, 41, 47, 59, 62, 64, 87, 90, 91, 116, 119

journal 41, 42, 43, 55, 56, 72, 73, 91, 103, 104, 105, 120, 122

knowledge-garbage 15, 19, 64

laboratories viii, 15, 33, 59, 62, 64, 69, 83
locality 3, 4, 8, 11, 20, 21, 22, 32, 96, 98, 99, 112

Machiavelli 86, 88, 89
machines 26, 62, 95, 116

management 26, 84, 103
material 6, 7, 8, 10, 11, 13, 14, 15, 31, 32, 48, 49, 50, 51, 52, 90, 97, 115, 117
ministers 33, 34, 38, 45, 50, 51, 78, 110
Modi, Narendra 33, 38, 41, 42, 43, 45, 51, 55, 56, 61, 91
municipalities 4, 6, 9, 11, 13, 24, 25, 26, 97, 98, 111

nature 24, 36, 37, 50, 56, 69, 78, 82, 83, 89, 98, 117, 123
neocolonial xvii, 34, 47, 73, 108
neoliberal xvi, xvii, 41, 46, 47, 48, 60, 62, 63, 66, 67, 68, 71, 72
nonhuman xi, xii
nurture 110

obstruction 79
ontology 85

pedagogy 61, 71
political party xvi, xvii, 4, 9, 33, 48, 49, 81, 109, 110
politics 3, 4, 5, 6, 7, 9, 10, 12, 13, 14, 15, 21, 25, 31, 32, 35, 37, 38, 39, 40, 45, 46, 47, 48, 49, 50, 54, 59, 60, 62, 67, 68, 70, 81, 84, 85, 87, 89, 98, 100, 107, 108, 109, 110, 115, 116, 117, 120
policy 12, 48, 78, 115
pollution 24, 46, 114
protest 4, 39, 68, 73, 90

queer 107

ragpickers 20, 26, 48, 104
religious 36, 45, 60, 81, 85, 110
residents ix, xvi, 3, 4, 11, 12, 19, 67, 82, 88, 96, 98, 104
rituals 10, 26, 31, 35, 36
roots 74, 76, 82, 90
rubbish 27, 42, 47, 56

sex 81
social 3, 4, 5, 6, 9, 10, 14, 21, 24, 34, 37, 39, 45, 47, 48, 50, 54, 60, 62, 67, 69, 84, 85, 87, 93, 95, 96, 98, 99, 107, 109, 110, 111, 112, 114, 116, 118, 119
spiritual 6, 7, 8, 9, 10, 11, 13, 34, 36, 49, 51, 70, 97, 108, 109
strategic 3, 4, 5, 25, 34, 38, 39, 40, 49, 51, 54, 78, 79, 81, 87, 112, 115, 117

townships 6
toxicity 32

urban 65, 87, 88, 96, 97, 99, 100, 101, 102, 103, 118

Varanasi 8, 9, 11, 12, 13, 17, 19, 22, 23, 24, 48, 93, 100, 108, 110, 112, 113
vision 36, 37, 40, 45, 46, 50, 60, 64, 78, 79, 88

waste vii, ix, xvi, 3–16, 19–28, 31, 32, 33, 36, 37, 39, 40, 41, 42, 44, 46, 47, 49, 50, 51, 54, 59–63, 65–73, 80, 85, 86, 87, 93–105, 110, 111, 112, 113, 117, 118, 119, 122
"worlding" 27, 32, 42, 46

zooplankton 3